T0222137

Veracity of Big Data

Machine Learning and Other Approaches to Verifying Truthfulness

Vishnu Pendyala

Apress®

Veracity of Big Data

Vishnu Pendyala
San Jose, California, USA

ISBN-13 (pbk): 978-1-4842-3632-1   ISBN-13 (electronic): 978-1-4842-3633-8
https://doi.org/10.1007/978-1-4842-3633-8

Library of Congress Control Number: 2018945464

Managing Director, Apress Media LLC: Welmoed Spahr
Acquisitions Editor: Celestin Suresh John
Development Editor: Laura Berendson
Coordinating Editor: Divya Modi

Cover designed by eStudioCalamar

Cover image designed by Freepik (www.freepik.com)

Distributed to the book trade worldwide by Springer Science+Business Media New York, 233 Spring Street, 6th Floor, New York, NY 10013. Phone 1-800-SPRINGER, fax (201) 348-4505, e-mail orders-ny@springer-sbm.com, or visit www.springeronline.com. Apress Media, LLC is a California LLC and the sole member (owner) is Springer Science + Business Media Finance Inc (SSBM Finance Inc). SSBM Finance Inc is a **Delaware** corporation.

For information on translations, please e-mail rights@apress.com, or visit http://www.apress.com/rights-permissions.

Apress titles may be purchased in bulk for academic, corporate, or promotional use. eBook versions and licenses are also available for most titles. For more information, reference our Print and eBook Bulk Sales web page at http://www.apress.com/bulk-sales.

Any source code or other supplementary material referenced by the author in this book is available to readers on GitHub via the book's product page, located at www.apress.com/978-1-4842-3632-1. For more detailed information, please visit http://www.apress.com/source-code.

Printed on acid-free paper

I dedicate this book to the loving memory of my father,
Pendyala Srinivasa Rao.

Table of Contents

About the Author

Vishnu Pendyala is a Senior Member of IEEE and of the Computer Society of India (CSI), with over two decades of software experience with industry leaders such as Cisco, Synopsys, Informix (now IBM), and Electronics Corporation of India Limited. He is on the executive council of CSI, Special Interest Group on Big Data Analytics, and is the founding editor of its flagship publication, *Visleshana*. He recently taught a short-term course on "Big Data Analytics for Humanitarian Causes," which was sponsored by the Ministry of Human Resources, Government of India under the GIAN scheme; and delivered multiple keynotes in IEEE-sponsored international conferences. Vishnu has been living and working in the Silicon Valley for over two decades. More about him at: https://www.linkedin.com/in/pendyala.

Acknowledgments

The story of this book starts with Celestin from Apress contacting me to write a book on a trending topic. My first thanks therefore go to Celestin. Thanks to the entire editorial team for pulling it all together with me – it turned out to be a much more extensive exercise than I expected, and the role you played greatly helped in the process.

Special thanks to the technical reviewer, Oystein, who provided excellent feedback and encouragement. In one of the emails, he wrote, "Just for your information, I learnt about CUSUM from your 4th chapter and tested it out for the audio-based motion detector that is running on our .... From metrics I could see that it worked really well, significantly better than the exponentially weighted moving average (EWMA) method, and it is now the default change detection algorithm for the motion detector in all our products!"

Introduction

Topics on Big Data are growing rapidly. From the first 3 V's that originally characterized Big Data, the industry now has identified 42 V's associated with Big Data. The list of how we characterize Big Data and what we can do with it will only grow with time. Veracity is often referred to as the 4th V of Big Data. The fact that it is the first V after the notion of Big Data emerged indicates how significant the topic of Veracity is to the evolution of Big Data. Indeed, the quality of data is fundamental to its use. We may build many advanced tools to harness Big Data, but if the quality of the data is not to the mark, the applications will not be of much use. Veracity is a foundation block of data and, in fact, the human civilization.

In spite of its significance striking at the roots of Big Data, the topic of its veracity has not been initiated sufficiently. A topic really starts its evolution when there is a printed book on it. Research papers and articles, the rigor in their process notwithstanding, can only help bring attention to a topic. But the study of a topic at an industrial scale starts when there is a book on it. It is sincerely hoped that this book initiates such a study on the topic of Veracity of Big Data.

The chapters cover topics that are important not only to the veracity of Big Data but to many other areas. The topics are introduced in such a way that anyone with interest in math and technology can understand, without needing the extensive background that some other books on the same topics often require. The matter for this book evolved from the various lectures, keynotes, and other invited talks that the author delivered over the last few years, so they are proven to be interesting and insightful to a live audience.

The book is particularly useful to managers and practitioners in the industry who want to get quick insights into the latest technology. The book has made its impact in the industry even before it was released, as the technical reviewer acknowledged in one of his emails that one of the techniques explained in this book that he reviewed, turned out to be better than the one they were using, so much so that the new technique from the book became the default for all the products in the product line.

The book can be used to introduce not only Veracity of Big Data, but also topics in Machine Learning; Formal Methods; Statistics; and the revolutionary technology, Blockchain, all in one book. It can serve as a suggested reading for graduate and undergraduate courses. The exercises at the end of each chapter are hoped to provoke critical thinking and stoke curiosity. The book can also be used by researchers, who can evaluate the applicability of the novel techniques presented in the book, for their own research. The use of some of the techniques described in the book for the problem of veracity are based on the author's own research.

The chapters can be read in any order, although there are some backward and forward references. Chapter 3, on the approaches to the Veracity of Big Data, gives a good overview of the remaining textbook and is a must for someone short on time. Blockchain is being touted as the ultimate truth machine that can solve a number of trust and veracity problems in the real world. The last chapter briefly introduces the topic of Blockchain as a trend that deserves to be keenly watched. It is sincerely hoped that the book will initiate a thorough study of the topic of veracity in the long run. Happy reading!

CHAPTER 1

The Big Data Phenomenon

We are inundated with data. Data from Twitter microblogs, YouTube and surveillance videos, Instagram pictures, SoundCloud audio, enterprise applications, and many other sources are part of our daily life. Computing has come of age to facilitate the pervasiveness of machine-readable data and leveraging it for the advancement of humanity. This Big Data phenomenon is the new information revolution that no IT professional can afford to miss to be part of. Big Data Analytics has proven to be a game changer in the way that businesses provide their services. Business models are getting better, operations are becoming intelligent, and revenue streams are growing.

Uncertainty is probably the biggest impeding factor in the economic evolution of mankind. Thankfully, Big Data helps to deal with uncertainty. The more we know an entity, the more we can learn about the entity and thereby reduce the uncertainty. For instance, analyzing the continuous data about customers' buying patterns is enabling stores to predict changes in demand and stock accordingly. Big Data is helping businesses to understand customers better so that they can be served better. Analyzing the consumer data from various sources, such Online Social Networks (OSN), usage of mobile apps, and purchase transaction records,

© Vishnu Pendyala 2018
V. Pendyala, *Veracity of Big Data*, https://doi.org/10.1007/978-1-4842-3633-8_1

businesses are able to personalize offerings. Computational statistics and Machine Learning algorithms are able to hypothesize patterns from Big Data to help achieve this personalization.

Web 2.0, which includes the OSN, is one significant source of Big Data. Another major contributor is the Internet of Things. The billions of devices connecting to the Internet generate Petabytes of data. It is a well-known fact that businesses collect as much data as they can about consumers – their preferences, purchase transactions, opinions, individual characteristics, browsing habits, and so on. Consumers themselves are generating substantial chunks of data in terms of reviews, ratings, direct feedback, video recordings, pictures and detailed documents of demos, troubleshooting, and tutorials to use the products and such, exploiting the expressiveness of the Web 2.0, thus contributing to the Big Data.

From the list of sources of data, it can be easily seen that it is relatively inexpensive to collect data. There are a number of other technology trends too that are fueling the Big Data phenomenon. High Availability systems and storage, drastically declining hardware costs, massive parallelism in task execution, high-speed networks, new computing paradigms such as cloud computing, high performance computing, innovations in Analytics and Machine Learning algorithms, new ways of storing unstructured data, and ubiquitous access to computing devices such as smartphones and laptops are all contributing to the Big Data revolution.

Human beings are intelligent because their brains are able to collect inputs from various sources, connect them, and analyze them to look for patterns. Big Data and the algorithms associated with it help achieve the same using compute power. Fusion of data from disparate sources can yield surprising insights into the entities involved. For instance, if there are plenty of instances of flu symptoms being reported on OSN from a particular geographical location and there is a surge in purchases of flu medication based on the credit card transactions in that area, it is quite likely that there is an onset of a flu outbreak. Given that Big Data makes no sense without the tools to collect, combine, and analyze data, some proponents even argue that Big Data is not really data, but a technology comprised of tools and techniques to extract value from huge sets of data.

Note   Generating value from Big Data can be thought of as comprising two major functions: *fusion,* the coming together of data from various sources; and *fission,* analyzing that data.

There is a huge amount of data pertaining to the human body and its health. Genomic data science is an academic specialization that is gaining increasing popularity. It helps in studying the disease mechanisms for better diagnosis and drug response. The algorithms used for analyzing Big Data are a game changer in genome research as well. This category of data is so huge that the famed international journal of science, *Nature,* carried a new item[1] about how the genome researchers are worried that the computing infrastructure may not cope with the increasing amount of data that their research generates.

Science has a lot to benefit from the developments in Big Data. Social scientists can leverage data from the OSN to identify both micro- and macrolevel details, such as any psychiatric conditions at an individual level or group dynamics at a macrolevel. The same data from OSN can also be used to detect medical emergencies and pandemics. In financial sector too, data from the stock markets, business news, and OSN can reveal valuable insights to help improve lending practices, set macroeconomic strategies, and avert recession.

There are various other uses of Big Data applications in a wide variety of areas. Housing and real estate business; actuaries; and government departments such as national security, defense, education, disease control,

---

[1]Hayden, Erika C.; "Genome Researchers Raise Alarm Over Big Data: Storing and Processing Genome Data Will Exceed the Computing Challenges of Running YouTube and Twitter, Biologists Warn," *Nature,* July 7, 2015, http://www.nature.com/news/genome-researchers-raise-alarm-over-big-data-1.17912

law enforcement, and energy, which are all characterized by huge amounts of data are expected to benefit from the Big Data phenomenon.

Note   Where there is humongous data and appropriate algorithms applied on the data, there is wealth, value, and prosperity.

Why "Big" Data

A common question that arises is this: "Why Big Data, why not just data?" For data to be useful, we need to be able to identify patterns and predict those patterns for future data that is yet to be known. A typical analogy is to predict the brand of rice in a bag based on a given sample. The rice in the bag is unknown to us. We are only given a sample from it and samples of known brands of rice. The known samples are called *training data* in the language of Machine Learning. The sample of unknown rice is the *test data*.

It is common sense that the larger the sample, the better the prediction of the brand. If we are given just two grains of rice of each brand in the training data, we may base our conclusion solely based on the characteristics of those two grains, missing out on other characteristics. In the Machine Learning parlance, this is called *overfitting*. If we have a bigger sample, we can recognize a number of features and a possible range of values for the features: in other words, the probability distributions of the values, and look for similar distributions in the data that is yet to be known. Hence the need for humongous data, Big Data, and not just data.

In fact, a number of algorithms that are popular with Big Data have been in existence for long. The *Naïve Bayes* technique, for instance, has been there since the 18th century and the *Support Vector Machine* model was invented in early 1960s. They gained prominence with the advent of the Big Data revolution for reasons explained earlier.

An often-cited heuristic to differentiate "Big Data" from the conventional bytes of data is that the Big Data is too big to fit into traditional Relational Database Management Systems (RDBMS). With the ambitious plan of the Internet of Things to connect every entity of the world to everything else, the conventional RDBMS will not be able to handle the data upsurge. In fact, Seagate[2] predicts that the world will not be able to cope with the storage needs in a couple of years. According to them, it is "harder to manufacture capacity than to generate data." It will be interesting to see if and how the storage industry will meet the capacity demands of the *Volume* from Big Data phenomenon, which brings us to the V's of the Big Data.

Note   It takes substantial data to see statistical patterns, general enough for large populations to emerge, and meaningful hypotheses generated from the data automatically.

The V's of Big Data

An Internet search for "President Trump" in double quotes returned about 45,600,000 results in just two weeks of his inauguration (January 20 – February 5). It is an indication of what have become known as the first four V's of Big Data: Volume, Velocity, Variety, and Veracity. The search results have all four ingredients. The corpus of 45M page results is the sheer *Volume* that has been indexed for the search. There could be many more on the topic. Given that the topic is just two weeks old, the results indicate

---

[2]Athow, Desire; "World Could 'Run Out of Storage Capacity' Within Two Years Warns Seagate: Blame the Data Capacity Gap," *TechradarPro*, December 22, 2014, http://www.techradar.com/news/internet/data-centre/world-could-run-out-of-storage-capacity-within-2-years-warns-seagate-vp-1278040/2

the speed or **Velocity** with which the data is getting created. Velocity also refers to the speed at which the data is analyzed and inferences from it put to use sometimes in real time, instantaneously.

The results exist in a wide **Variety** of media: text, videos, images, and so on – some of it structured, some of it not – and on a wide variety of aspects of President Trump. How much of that information is true and how many of those results are about "President Trump" and not the presidential candidate Trump? A quick scan shows that quite a few results date back to 2016, indicating that the articles are not really about "President Trump" but may be about "the presidential candidate Trump." That brings us to the problem of **Veracity** of Big Data. Veracity refers to the quality of data. How much can we rely on the Big Data? Initially, the definition of Big Data included only the first three V's. Visionary corporates like IBM later extended the definition to include veracity as an important aspect of Big Data.

There are more V's that are increasingly being associated with Big Data. Elder Research came up with 42 V's associated with Big Data and Data Science. While we will not go over all of them here, we will cover the important ones and introduce a new one, Valuation, to the mix. The next V is Variability. The meaning of data can vary depending on the context. For instance, an error of 0.01 inch when measuring the height of a person is very different from the same error when measuring the distance between the two eye pupils in a face recognition algorithm. The 0.01" error datum has varied meanings in the two different contexts, even if both are used in the same application to recognize people. This is often true of spoken data as well. Interpretation of speech greatly depends on the context. This notion is called the Variability of Big Data. Big Data Analytics therefore needs to be cognizant of the **Variability** of Big Data, when interpreting it.

Variability presents a formidable challenge for deriving value from the Big Data. Data, by itself is useless. We need to derive **Value** from it by analyzing it and drawing conclusions or predicting outcomes. For executives to see the value of Big Data, it needs to be presented in a way they understand. ***Visualization*** is the representation using diagrams, charts, and projections of the Big Data in ways that are more understandable, in order for its value to be realized. It is believed that 95% of the value from Big Data is derived from 5% of its attributes or characteristics. So, only 5% of the attributes are really viable. ***Viability*** refers to the fact that not everything about Big Data is useful. There is a need to evaluate the viability of available attributes to choose only those, which help getting the value out.

The V's enunciated so far miss out on a very important aspect of Big Data – extracting value from the data using algorithms. We call it ***Valuation***, thereby introducing a new V, probably the most important of all. Without the process of valuation, the data is useless. The Data Mining, Machine Learning, and other methods from computational statistics, comprise the tool set for Big Data valuation. The efficiency and effectiveness of this tool set is critical to the success of the Big Data phenomenon. Valuation is not to be confused with Value. The former is a process and the latter is the output. Valuation is the process of extracting value from Big Data.

Note   Big Data is often associated with V's: Volume, Velocity, Variety, Veracity, Value, Variability, Visualization, and Viability. We introduce a new V – Valuation – the process of extracting value from Big Data using algorithms.

The various V's are all related (Figure 1-1). If we take the case of Social Media, it can be generally said that its veracity is inversely proportional to the velocity. The faster blogs and microblogs are posted, the lesser the thought and substantiation behind them. When videos were first posted for watching online, the volume of the Big Data grew enormously and at a great speed. Variety therefore contributed to the volume and velocity of the data generated.

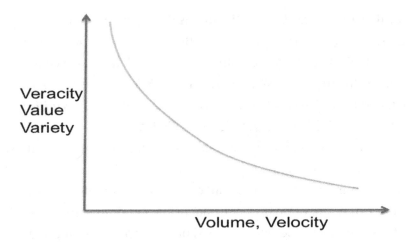

Figure 1-1. *Relationship between the V's of Big Data*

Variety also contributes to the veracity of information. For instance, a social media post that includes a video or a picture is more likely to be true than the one with just text. Volume is inversely related to veracity because more data points may mean lesser accuracy. In case of information and data, noise increases with the signal. In some cases, noise actually increases faster than the signal, impacting the utility of the data. Falsely crying "Fire" in a movie hall spreads faster than the true opinion that the movie is not worth watching. The same applies to information on the Web, when noise travels faster than the signal. Hence, value is directly proportional to veracity and inversely related to volume and velocity. However, high-quality data does not automatically guarantee high value. We also need efficient algorithms used for the valuation process. All these ideas are summarized in Figure 1-1.

From the above characteristics described in terms of V's, it must be clear now that Big Data is quite different from the conventional schemes of data that usually resided in databases and other structured frameworks. The major differences are that the former is mostly unstructured, raw, and real time. The traditional tools of data warehousing, business process management, and business intelligence fall grossly short when it comes to handling the all-encompassing Big Data. A number of decades old techniques, mostly discovered as part of Artificial Intelligence and Machine Learning, have now become relevant in the context of Big Data. These tools, techniques, and algorithms are able to deal with the unique characteristics of Big Data.

Note   The V's, which characterize Big Data, make it hard for the traditional ETL (Extract, Transform, Load) functions to scale. ETL methods can neither cope with the velocity of the data generation nor can deal with the veracity issues of the data. Hence there is a need for Big Data Analytics.

Veracity – The Fourth 'V'

We have seen a number of uses of Big Data in the preceding sections. All these applications rely on the underlying quality of data. A model built on poor data can adversely impact its usefulness. Unfortunately, one of the impediments to the Big Data revolution is the quality of the data itself. We mentioned earlier how Big Data helps in dealing with uncertainty. However, Big Data itself is characterized by uncertainty. Not all of Big Data is entirely true to be able to deal with it with certainty. When the ground truth is not reliable, even the best-quality model built on top of it will not be able to perform well. That is the problem of Veracity, the fourth 'V' of Big Data. Veracity is a crucial aspect of making sense out of the Big Data and getting *Value* out of it.

Value is often touted as the fifth 'V' of Big Data. Poor quality of data will result in poor quality of analysis and value from it, following the "Garbage in, garbage out" maxim. There have been significant failures because veracity issues were not properly handled. The Google Flu Trends (GFT)[3] fiasco is one. Google's algorithms missed on quite a few aspects of veracity in drawing their conclusions and the predictions were off by 140%. If the healthcare industry makes preparations based on such predictions, they are bound to incur substantial losses.

Note   Veracity of Big Data can be defined as the underlying accuracy or lack thereof, of the data in consideration, specifically impacting the ability to derive actionable insights and value out of it.

There are many causes for poor data quality. Uncertainty in the real world, deliberate falsity, oversight of parts of the domain, missing values, human bias, imprecise measurements, processing errors, and hacking all contribute to lowering the quality. Figure 1-2 shows the factors influencing the veracity of data. Sensor data from the Internet of Things is often impacted by poor measurements and missed out values. Of late, OSN have been targets of increasing episodes of hacking. Uncertainty in the real world is unavoidable. Rather than ignore it, we need methods to model and deal with uncertainty. Machine Learning offers such methods.

---

[3]Lazer, David, and Kennedy, Ryan; "What We Can Learn from the Epic Failure of Google Flu Trends," October 1, 2015, *Wired*, https://www.wired.com/2015/10/can-learn-epic-failure-google-flu-trends/

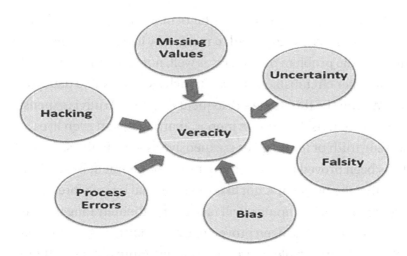

Figure 1-2. *Factors influencing the veracity of data*

For effective models and methods using them, the dataset needs to be representative of and present a wholesome picture of the domain. For instance, governments, particularly those of the developing countries such as India, are increasingly depending on microblogging websites such as Twitter to make their decisions. While this may work in quite a few cases, the dataset is not a wholesome representation of the government's target population because it excludes a substantial chunk of the citizens who have no presence in social media for various reasons. Models built on less representative datasets are bound to fail.

Note   Choice of the dataset is critical to the veracity factor and success of a Big Data model. Datasets must be comprehensive representations of application domains; and devoid of biases, errors, and tampered data points.

Machine Learning is the new mortar of modernization. Machine Learning algorithms are now used in many new inventions. They can be used to solve the problem of Veracity as well, not just for the uncertainty case but in other circumstances as well, such as when some data points are missing. We will see this in the subsequent chapters. Truth is expensive – history has taught us that it costs money, marriages, and even lives. Establishing truth or limiting lies is expensive as well. In the software world too, it has been proved that Misinformation Containment is NP-Hard, implying that the problem is not likely to be solved by software running in polynomial time. Computers will take an inordinately long time to solve a problem that has been proven to be NP-Hard, making the solution infeasible. NP-Hard problems are solved using approximate methods. Machine Learning offers algorithms that help in solving the problem of Misinformation Containment approximately.

Note   Misinformation Containment (MC) is NP-Hard, unlikely to be solved accurately by algorithms completing in reasonable time.

Microblogs are a significant chunk of the Big Data, which is substantially impacted by the problem of poor quality and unreliability. According to Business Insider,[4] it is estimated that in the month of August 2014, on average, 661 million microblogs were posted in a single day on Twitter alone. The amount of care and caution that would have gone into that many tweets posted at that speed is anyone's guess. Consumers rather than businesses mostly build the digital universe. It implies that most of the data in the digital universe is not sufficiently validated and does not come with the backing of an establishment like for the enterprise data.

---

[4]Edwards, Jim; "Leaked Twitter API Data Shows the Number of Tweets Is in Serious Decline," *Business Insider*, February 2, 2016, http://www.businessinsider.com/tweets-on-twitter-is-in-serious-decline-2016-2

A significant portion of the data supposed to be from businesses is not entirely true either. A simple Internet search for "super shuttle coupon" on February 17, 2017, returned, on the very first page, a result from http://www.couponokay.com/us/super-shuttle-coupon announcing 77% off, but clicking the link returns a vague page. We know shuttle companies cannot afford to give 77% off, but the Internet search still pulls up the URL for display in the results. It may have been so much better if the search engines quantified their belief in the results displayed and eliminated at least the blatantly false ones.

The readers will also probably be able to relate their own experiences with the problem of veracity of enterprise-backed data and information. We have all been inconvenienced at one time or other by misleading reviews on Yelp, Amazon, and the websites of other stores. The issue actually impacts the businesses more than consumers. If a product receives many positive reviews incorrectly, businesses like Amazon may stock more of that product and when the truth dawns on the customers, may not be able to sell them as much, incurring losses.

In general, however, as Figure 1-3 shows, the quality of the data generated deteriorates from enterprise to sensor. Machine data, such as from the logs, clickstream, core dumps, and network traces, is usually quite accurate. It can be assumed that healthcare data has gone through the due diligence that is necessary to save lives, so it can be ranked relatively high in terms of accuracy. Static Web content is associated with the reputation and accountability of a web domain, so it can be assumed to be reasonably accurate. Social media is still casual in nature, with open access and limited authentication schemes, so it suffers in terms of quality. Sensors operate in low energy, lossy, and noisy environments. Readings from sensors are therefore often inaccurate or incomplete.

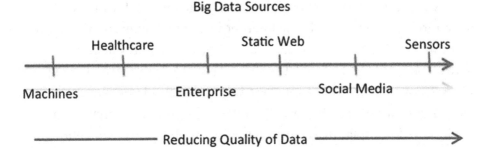

Figure 1-3. *Big Data sources listed in decreasing order of veracity*

However, sensor data can be crucial at times. Take for instance, the case of the fatal accident involving the Tesla Model S car.[5] According to the reports in the press, the accident is caused because the data from the radar sensor did not align with the inference from the computer vision-based vehicle detection system. Therefore, the crash-avoidance system did not engage and a life was lost. While sensor readings can be improved by using better hardware, there is also a need for algorithms to accommodate the poor quality of data from the sensors, so that disasters such as the car accident can be avoided.

The first scientific step to model complexity in the world has often been to express it in math. Math helps to improve the understanding of matter. We therefore use math where possible to enhance our comprehension of the subject. Algorithms can then be derived from the math, where needed.

Note   Math is meant to make matters easy, not tough.

---

[5]Singhvi, Anjali, and Russell, Karl; "Inside the Self-Driving Tesla Fatal Accident," *New York Times*, July 12, 2016, https://www.nytimes.com/interactive/2016/07/01/business/inside-tesla-accident.html

Summary

Big Data is changing the world in substantial ways. In this chapter, we examined this phenomenon and covered important ideas pertaining to it. We reviewed the various V's that apply to Big Data and introduced a new 'V', Valuation. Veracity is an important V. Social media is toward the lower end of the Veracity spectrum. In the next chapter, we discuss this issue more closely, looking at the some of the ramifications of inaccuracy in the information posted on social media and the Web in general.

EXERCISES

1. Draw a concept map of the various ideas presented in the chapter.

2. Think of any other V's that apply to Big Data. Discuss their potential.

CHAPTER 2

Veracity of Web Information

The problem of the Veracity of Big Data can best be understood by one big source of Big Data – the Web. With billions of indexed pages and exabytes in size, Web Information constitutes a substantial part of the Big Data. It plays a crucial role in affecting people's opinions, decisions, and behaviors in their day-to-day lives. The Web probably provides the most powerful publication medium at the lowest cost. Anyone can become a journalist just with access to the Internet. This is good, because it brought a number of truths and a number of otherwise unnoticed people to the forefront. However, the same reasons that make the Web powerful also make it vulnerable to malicious activity. A substantial portion of the Web is manifested with false information. The openness of the Web makes it difficult to sometimes even precisely identify the source of the information on it.

Electronic media has been used to mislead people even in the past. On October 30, 1938, a 62-minute dramatization on the radio, of H. G. Wells's science fiction, "The War of the Worlds" with real-sounding reports terrified many in New York, who believed that the city was under attack by aliens. Information in the age of the Web travels even wider and faster. In microseconds, even remote corners of the world can be alerted. Information travels at almost the speed of light and so will lies and the tendency to tell lies. The unrivaled reach and reliance on the Web

© Vishnu Pendyala 2018

V. Pendyala, *Veracity of Big Data*, https://doi.org/10.1007/978-1-4842-3633-8_2

is increasingly being used to perpetuate lies owing to the anonymity and impersonal interface that the Web provides.

We have seen a number of people rising to fame using the social media. Many a tweet has become a hotly discussed news item in the recent past. The Web is unique in allowing a few people to influence, manipulate, and monitor the minds of many. Hashtags, likes, analytics, and other artifacts help in monitoring. The feedback from these artifacts can then be used to fine-tune the influence and manipulate the people. The Web indeed provides an excellent tool for fostering the common good. But the same power of the Web can be used against the common good to favor a select few, who have learned the art of manipulation on the Web. In this chapter we will examine the problem of veracity of the Web information: its effects, causes, remedies, and identifying characteristics.

Note   The Web is a substantial contributor to the Big Data and is particularly prone to veracity issues.

The Problem

When an earthquake struck near Richmond, Virginia, in the United States, people living about 350 miles away in New York City, read about it on the popular microblogging website, Twitter, 30 seconds before the earthquake was actually felt in their place. Governments use Twitter to locate their citizens in times of need to offer services. The Web prominently figures in multiple governments' strategy for interacting with the people and manage perceptions. This book itself relies to some extent on the veracity of the Web's information, as can be seen from the footnote citations. People the world over increasingly depend on the Web for their day-to-day activities such as finding directions to places on a map, choosing a restaurant based on the reviews online, or basing opinions on the trending topics by reviewing microblogs. All these activities and more can be adversely impacted by inaccuracy of the content that is relied upon.

Lies in general have a substantial cost and lies on the Web have that cost multiplied manifold. For instance, in January 2013 [1], false tweets that a couple of companies were being investigated cost the respective companies 28% and 16% of their market capitalization – a huge loss to the investors. These are blatant attempts. There are also subtle ways that the Web has been used to manipulate opinions and perceptions. Chapter 4, "Change Detection Techniques" discusses the case of how perception manipulation on the Web was successfully used to change the outcomes of presidential elections in South American countries. Cognitive hacking refers to this manipulation of perception for ulterior motives. Cognitive hacking is a serious threat to the veracity of Web information because it causes an incorrect change in behavior of the users. On April 10, 2018, in his testimony to the US congress, Facebook's Mark Zuckerberg acknowledged not doing enough to prevent foreign interference in US elections. Such is the role Social Media and cognitive hacking play in today's governance.

All information posted on the Web with ulterior motives, conflicting with the accepted values of humanity, can be considered as having compromised veracity. There are many website schemes that fit this description. The popular ones are in the stock exchange realm, where monetary stakes are high. The US Securities and Exchange Commission (SEC) has issued several alerts and information bulletins[2] to detail the fraudulent schemes and protect the investors. There are other subtle ways that integrity of the Web is compromised. In spamdexing or "black hat

---

[1]US Securities and Exchange Commission; SEC Charges: False Tweets Sent Two Stocks Reeling in Market Manipulation, November 5, 2015, Press Release, https://www.sec.gov/news/pressrelease/2015-254.html

[2]US Securities and Exchange Commission; SEC Charges: Internet Fraud, February 1, 2011, Investor Publication, https://www.sec.gov/reportspubs/investor-publications/investorpubscyberfraudhtm.html

search engine optimization," Web search engines are tricked in a number of ways to show incorrect results of the search queries. Some of these tricks include posting unrelated text matching the background color, so that it is visible to the search engines but not to the users and hiding the unrelated text in HTML code itself using empty DIVs.

The problem manifests in a number of forms. Almost every useful artifact has a potential for misuse. For instance, quite a few websites such as LinkedIn and Twitter shorten long URLs in the user posts, for ease of use. There are websites like Tinyurl.com, which provide this service of shortening the URL, for free. The shortened URL often has no indication of the target page, a feature that has been misused to make gullible users to visit websites they would not have otherwise visited. The shortened URL is sent in emails with misleading but attractive captions that tend to make the recipients click on it, landing them on obscure websites. The Pew Research Center estimated that 66% of the URLs shared on Twitter come from bots, which are software programs that run on the Internet to automatically perform tasks that a human being is normally expected to do.

Along these lines is the technique of cloaked URL, which appears genuine, but may either have hidden control characters or use domain forwarding to conceal the actual address of the website to which the URL opens up. Spoofing websites in these and other ways has been going on since almost the inception of the Web. Miscreants create websites that look exactly like the original authentic websites and have a similar URL, but with misleading content and fake news. A more serious form of spoofing is phishing, where the users are misled into sharing their sensitive information such as passwords, social security numbers, and credit card details when they visit a fake website that looks almost identical to the corresponding genuine website.

Note   The problem of fraud on the Web takes many forms. The problem is complex from a technical and sociological perspective.

The Causes

Web 2.0 generated several petabytes of data from people all over the world, so much so that today, the digital universe is mostly built by consumers and ordinary users rather than businesses. Consumer-generated data has grown manifold from the time the 2.0 avatar of the Web came into existence. It implies that most of the data is not validated and does not come with the backing of an establishment like that of the business data. The Web's role as the single source of truth in many instances has been misused to serve hidden agendas. In spite of its anthropomorphic role, unlike human beings, the Web does not have a conscience. Still there is often more reliance on the Web than on the spoken word.

Like with many inventions such as atomic energy, when envisioning the Web, the euphoria of its anticipated appropriate use seems to have dominated the caution to prevent misuse. On one hand, indeed, the Web may not have grown as it did today if it was restrictive to permit only authenticated, genuine, and high-quality content. On the other hand, the Web's permissiveness is causing a complete lack of control on the huge chunk of the Big Data that the Web contributes. The original purpose of creating the Web was apparently to encourage peaceful, mutually beneficial communication among the people the world over. However, the loose controls enabled not just people, but also swarms of masquerading bots, to automatically post misleading information on the Web for malicious motives.

When the author presented his work on determining truth in Web microblogs in a conference, a research student in the audience expressed the opinion that the Web is a casual media for information exchange and should not be judged for truthfulness. Insisting on a truthful Web will make it less interesting and far less used, was her rationalization. It is a reflection on the perception of a substantial number of contributors of the Web content. The casual attitude of the users posting information is one of the major causes for the lack of veracity of Web information.

The gray line between permitting constructive content to be posted and restricting abusive information is hard to discern. Policing malicious online conduct is certainly not easy.

There is plenty of incentive for perpetuating fraud on the Web. As we shall see in a later chapter, even the presidential elections of a country were impacted by the information posted on the Web, and at times, incorrectly. The biggest impact of public opinion is probably on the politicians. Political agenda is probably the most significant cause of manipulation using the Web. The next motivating factor is money, as we saw in the case of stock price manipulation. Monetary rewards also accrue in the form of advertising on spoofed and fake news websites. Revenge and public shaming is also a common cause, so much so that a majority of the states in the United States have passed laws against posting revenge porn online.

Cognitive hacking is particularly effective on the Web because the interaction lacks a number of components that help in gauging the veracity of the spoken word. For instance, if the speaker is in front, the genuineness of the words is reflected in body language, eye contact, tone, and the continuous feedback loop between the speaker and the listener. This is illustrated in Figure 2-1. As lives become increasingly individualistic, people depend more and more on the Web for information, missing some of the veracity components shown in the bidirectional arrow in Figure 2-1, which are present in a live conversation. There are many cases where websites have been used to cash on this situation to mislead the public for ulterior motives, often illegally.

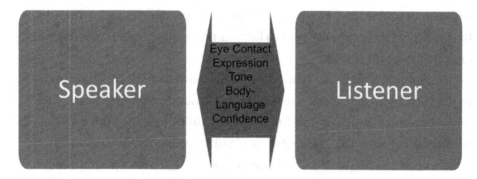

Figure 2-1. *Communication on the Web does not have some of the veracity components or the feedback loop of a live conversation*

In an increasing trend, software is being written not only to do what is told, but also to craft its own course of action from the training data and a preferred outcome. This trend also is contributing to the growing amount of fake data. For instance, speech synthesis software is now capable of impersonating people's voices. Lyrebird[3], a company which makes such software, allows one to record a minute from a person's voice. Using the recording, the company's software generates a unique key, something like the DNA of a person's voice. Using this key, the software can then generate any speech with the person's voice.

Stanford University's Face2Face[4] can take a person's face and re-render a video of a different person to make it appear as if the video is that of this new person. The software replaces the original face with the obtained one and generates the facial expressions of the faked face, making it appear as if the person in the video is the new person. Interestingly, projects like these are started to actually detect video manipulation, but end up being used more for actually manipulating the videos. Combined with speech-synthesizing

---

[3]Lyrebird, Beta Version, Generate sentences in your own voice, Last Retrieved: April 6, 2018, https://lyrebird.ai/

[4]Stanford University, Face2Face: Real-time Face Capture and Reenactment of RGB Videos, Last Retrieved: April 6, 2018, http://www.graphics.stanford.edu/~niessner/papers/2016/1facetoface/thies2016face.pdf

software described in the above paragraph, it is possible to make the videos look genuine. It is probably not too far in time that fake videos are generated entirely new from scratch.

Generating fraudulent content is inexpensive as well, contributing to the veracity issues. There are multiple businesses, which generate fake news and indulge in fraudulent manipulations on the Web for an alarmingly low cost. A Chinese marketing company, Xiezuobang, charges just $15 for an 800-word fake news article. There are a number of companies that offer similar fraudulent services to manipulate the data from the Web. Buying followers, +1's, and likes on social media; inappropriately boosting a website's or video's rank in search results; and fraudulent clicking on "pay-per-click" advertisements is not only possible, but also affordable. In fact, the cost of faking on the Web is often far less than advertising genuine content. Some of the malicious hacking, such as "Click Fraud," impacting the clickstream data, which is an important component of the Big Data, has been reined-in by use of technology. But newer, affordable mechanisms to dismantle the integrity of Big Data keep springing up from time to time.

Note   It is much easier to mislead people on the Web than in a face-to-face communication.

The Effects

The impact of fraud on the Web has been grievous and even fatal in some cases. There is extensive literature on the ramifications of the falsity on the Web. Hard to believe, but the lenience of the Web infrastructure has permitted websites that advertise criminal services such as murder and maiming. Such services often are based on lies and fraudulent information. Even the most popular websites such Craigslist and Facebook were used to seek hitmen. In one instance, a post on Facebook read, "... this girl knocked off right now" for $500 [5]. There are many untold stories abound about much worse scenarios from the "Dark Web," which operate on the Internet, but discretely, requiring special software, setup, and authorizations for access.

The effects of fraud on the Web varies with the schemes we discussed earlier. For instance, popular among the stock price manipulation schemes is the "Pump-and-Dump" pattern where the stock price is first "pumped" by posting misleading, incorrect, and upbeat information about the company on the Web and then "dumping" the miscreants' stocks into the market at an overpriced value. The hype affects many investors and the company in the long run. However, the impact is probably felt most when the Web is used for anti-democratic activities, manipulating opinions to the tune of changing the outcomes of countries' presidential elections.

Similar to the website spoofing we discussed earlier, Typosquatting or URL hijacking is when a typo in the URL of a reputed website leads to an entirely different website that displays annoying advertisements or maliciously impacts the user in some other ways. Whitehouse.com, even today does not seem to do anything with the home of the president of the United States, but instead was used to advertise unrelated stuff. At least now, the website includes a fineprint that they "are not affiliated or endorsed by U.S. Government," which was not the case a year ago. If users wanting to type whitehouse.gov instead type whitehouse.com, they will land up on this unexpected website, which claims to be celebrating its 20 years of existence. There are also cases of cybersquatting when domains were acquired with bad intentions. For instance, the domain, madonna.com was acquired by a person unrelated to the famous pop singer, Madonna.

---

[5]Black, Caroline; Facebook Murder-for-Hire Plot: Pa. Teen Corey Adams Admits He Used Site to Go After Rape Accuser, *CBS News*, February 14, 2011, http://www.cbsnews.com/news/facebook-murder-for-hire-plot-pa-teen-corey-adams-admits-he-used-site-to-go-after-rape-accuser/

The website was used to display porn instead of information about the singer. The impact in such cases ranges from a mere annoyance to serious loss of reputation.

Back in 1995 PETA, which stands for "People for the Ethical Treatment of Animals," was shamed by peta.org, which incorrectly talked about "People Eating Tasty Animals." We continue to hear about how reputed websites get hacked and defaced. A group called "Lizard Squad" defaced the home pages of a number of companies, including Lenovo and Google (in Vietnamese) in the past. At one time, the United Kingdom's *Vogue* magazine's website was defaced to show dinosaurs wearing hats. Quite a few times, political websites, such as that of the militant group, Hamas in 2001, was hacked to redirect to a porn website. Even Internet companies, which wield a substantial control over the Web were not immune to these attacks. It was mentioned earlier that Google's Vietnamese home page was defaced by the "Lizard Squad." In 2013 and again in 2014, millions of Yahoo's user accounts were hacked, impacting the company's net worth adversely.

In its "State of Website Security in 2016" report,[6] Google has warned that the number of hacked websites has increased by 32%, which is quite alarming. The hackers are becoming more ingenious with time and deploying advanced technologies to perpetrate crime. There is the case of hackers using 3D rendering on Facebook photos to try to trick the authentication systems using facial recognition to gain entry. Mobile Virtual Reality is capable of providing from a 2D picture of someone's face, the detail in 3D that is needed by the authentication systems. As we know, breach of security of systems has significant consequences.

Inaccuracy on the Web has a ripple effect into the applications, which rely on the veracity of the Web. The problem gets compounded as more and more things from the Internet of Things space start leveraging the Web. If the devices connecting to the Web are not secure enough, bots can be used to attack websites in various ways or even bring them down.

---

[6]Google, #NoHacked: A year in review, *Webmaster Central Blog*, March 20, 2017, https://webmasters.googleblog.com/2017/03/nohacked-year-in-review.html

Note   The Web today and the falsehood thereof, is capable of changing the trajectory of nations and the lives of billions.

The Remedies

The cause and effect of the problem of veracity of Big Data form a vicious cycle, as shown in Figure 2-2, one feeding into the other. An effective remedy needs to break this vicious cycle. Extinguishing the causes or diminishing the effects or both can help in containing the problem. Disincentivizing, punishing, and preventing the occurrence are some good strategies that various entities fighting Web-based fraud have used.

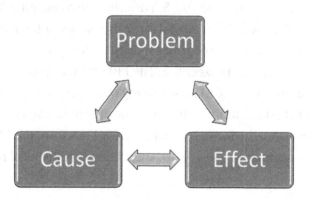

Figure 2-2. *The Cause-Effect Vicious Cycle of the Veracity Problem*

Given the repercussions and ramifications of fraud on the Web, several governments have made it a national priority to address the abuse. Governments like that of Russia maintain websites that contain fake news.[7] Interestingly, the European Union too reviews and posts

---

[7]Ministry of Foreign Affairs of the Russian Federation, Examples of publications that duplicate unreliable information about Russia, Press Release, Last Retrieved: April 6, 2018, http://www.mid.ru/nedostovernie-publikacii

disinformation,[8] but that which is pro-Russian government. There are also a number of independent organizations and websites like snopes. com that are constantly involved in debunking false information floating on the Web and the Internet in general. Social media providers too are actively engaged in curbing falsity on their sites. Suspension of fake accounts, awareness campaigns, software enhancements to flag fraud in the newsfeeds, and allowing users to provide credibility ratings are some of the experiments that the Internet companies are trying. Governments have been passing laws against online crime from time to time.

A number of crimes highlighted in the previous sections are covered by various sections of the law of the land. For instance, the person posting false tweets about the two companies, causing their stock prices to go down substantially, was charged under Section 10(b) of the Securities Exchange Act of 1934 and Rule 10b-5. The crime also caused SEC to issue an investor alert[9] in November 2015. However, the law or the alerts haven't caused the Web fraud in the securities market to cease. New cases are still being unearthed and prosecuted. The FBI set up a form at its Internet Crime Complaint Center (IC3) for reporting online crimes[10] and publishes an annual report on the topic.[11] However, the author himself reported a grievous crime using the form, but did not hear back from the FBI or even get a reference number for the report, suggesting that the legal remedies

---

[8]European Union, Disinformation Review, External Action Service, September 2, 2016, https://eeas.europa.eu/headquarters/ headquarters-homepage_en/9443/Disinformation%20Review

[9]US Securities and Exchange Commission, Social Media and Investing – Stock Rumors, Investor Alerts and Bulletins, November 5, 2015, https://www.sec.gov/ oiea/investor-alerts-bulletins/ia_rumors.html

[10]US FBI, File a Complaint, Internet Crime Complaint Center, Last Retrieved: April 6, 2018, https://www.ic3.gov/complaint/default.aspx

[11]US Department of Justice, 2015 Internet Crime Report, Internet Crime Complaint Center, Last Retrieved: April 6, 2018, https://pdf.ic3.gov/2015_IC3Report.pdf

are inadequate. Moreover, law does not even exist or is too easy on a lot of mischief played on the Web. It is imperative that the problem be addressed using techniques from not just legal, but other disciplines, particularly from technology.

Technology itself played an important role in resolving several legal mysteries, but it is far from getting the needed credibility. A simple application of the Bayes' theorem played a major role in a number of legal cases, some of which are listed at https://sites.google.com/site/ bayeslegal/legal-cases-relevant-to-bayes. As of date, the website lists 81 such cases. However, as the multiple books and the Internet search for the string, "and to express the probability of some event having happened in percentage terms" reveal, a British court of appeals banned the use of Bayes' theorem. There is a need to instill confidence in technological solutions to the problem of veracity, by strengthening the approaches.

Note   A simple application of the Bayes' theorem learned in high school math helped solve the mysteries of numerous legal cases.

Privacy protections make prosecuting an online crime difficult, particularly when the crime occurs across the countries' borders. Even if prosecution succeeds, the damage done is often irreparable and the time and matter lost is irretrievable. The culprits often do not have the resources to pay the damages, such as in case of the loss to a company's net worth owing to a few individuals' "Pump-and-Dump" activity. Hence, prevention is more important than punishment. Technology is far better at prevention than legal remedies. The subsequent chapters delve more into the technologies that can help in preventing online fraud.

Current technology also needs to be rechristened to address the veracity concerns of data. As an example, by prioritizing popularity over trustworthiness in showing the search results, search algorithms like PageRank are ignoring the problem. Ranking information retrieved based on its trustworthiness is a much more difficult task than recursively computing the number of links to a page, which the current search algorithms focus on. Internet companies such as Facebook, Google, and Twitter have been coming up with ways to counter fake and malicious content. Google, in the fourth quarter of 2016, for instance, banned 200 companies in its ad network for their misleading content. Facebook, in the first 90 days of 2018, deleted 583 million fake accounts and 865.8 million posts in addition to blocking millions of attempts to create fake accounts every day.

The real world has been a source of many engineering solutions. Quite a few concepts in Computer Science, such as the Object-Oriented paradigm, are inspired by real-world phenomenon. Veracity is a major problem in the real world too. How is it tackled there? How do the law enforcement officers know if a person is telling the truth? It is with the experience of examining a number of people involved in a variety of cases that they are able to make out the truth. A similar concept in the software world is Machine Learning. By examining patterns in a number of cases (training data), Machine Learning algorithms are able to come to conclusions. One of the conclusions can be whether the data is true or false. We will examine more details about how the Machine Learning techniques are used to solve the Big Data veracity problem.

Biometric identification, two-factor authentication, strong passwords, and such basic security features may be able to prevent data breaches, hacking, and information theft. But as we saw, like thieves that are always one step ahead, hackers are much more ingenious in their ways. Can we develop a Central Intelligence Agency (CIA)-like platform to detect the threats from hackers? The Threat Intelligence Platform (TIP) comes close to this. Just like the CIA, the TIP envisages to gathering threat information from various sources, correlates and analyzes this information, and acts based on the analysis. The field is still emerging and companies such as Lockheed Martin are working on robust solutions.

Note   The real world has been a source of inspiration for many engineering solutions, including to the problem of veracity of Big Data.

In spite of the vulnerability of the Web, a tiny segment of it is reasonably maintaining its integrity. Wikipedia pages are relied upon by not only individuals, but also by search companies. In spite of allowing anonymity, barring a few incidents of conflicted and "revenge" editing, Wikipedia has immensely benefited from its contributors in maintaining the integrity and high standards of the content. Unlike the real-world example of too many cooks spoiling the broth, collaborated editing has surprisingly proven to be highly successful. It is yet to be seen if collaborated editing is the silver bullet remedy to the problem of veracity, but in general what makes a website trustworthy? In the next section, we will examine the ingredients of the websites, which look relatively more reliable.

Characteristics of a Trusted Website

Trusted websites have certain characteristic features. A trusted website is analogous to a trusted person. We trust a person when he is objective, devoid of any hidden agenda or ulterior motive, represents an established company, speaks in clear and formal sentences, demonstrates expertise, is helpful, is recommended by trustworthy people, easily accessible, appears formal and real in general, and keeps himself abreast with latest developments, to name a few. The same goes for websites. We normally do not suspect weather websites because the information is objective. There is hardly any motive to manipulate the weather information. The information is simple and easy to interpret. It is not hard to see if the information is real or not because the temperature or humidity can be readily felt and contrasted with what is posted on the webpage. Any predictions can also be historically correlated with real happenings. All weather websites worth their name will keep them updated often, so as to stay in business.

Correlation of the information to established facts is also a good indication of veracity. For instance, if the webpage has a number of journal articles referenced, it is likely to be more trustworthy. The source of a website, such as the government or a reputed publisher also is a factor in determining the credibility of a website. The tone, style, clarity, and formality of the language used are also good indicators of the veracity of the information. A number of advertisements appearing on the webpage, particularly the annoying ones popping up windows, blocking content, or forcing a mouse click, make the website less reliable. Some of these advertisements can even inject spyware or viruses.

Authentic websites tend to include contact information such as a phone number and an email address. Factors such as appearance, consistency, quality of language used, community involvement and support, intelligibility, succinctness, granularity, well-organized and layered structure, currency, and usefulness of the information also contribute to the veracity of the information presented on a website. On the other hand, absence of any individual names or contact information, misspellings, unprofessional look and layout, vague references, intrusive advertisements, and such are tell-tale signs of a fraudulent website. We shall see how some of these and other features help with the solutions to the veracity problem in later chapters.

Note   A trusted website often exhibits many characteristics of a trusted person.

Summary

In this chapter, we discussed the cause-effect vicious circle of the veracity problem and looked at a few possible remedies. The next chapter will further explore the remedies from a technical perspective and discuss general approaches for dealing with the veracity issues of information. This chapter also reviewed the features of reliable websites. Some of these features like the presence of valid contact information, broken links, or popularity in terms of indices like PageRank can be automatically detected by software programs. However, features like the objectivity or completeness of the information or how updated they are, can be affirmed only with human involvement. The next chapter examines some of the ways of establishing credibility of information.

EXERCISES

1. List out as many of the factors that affect trustworthiness in the real world. Examine which of those apply to the Veracity of Web Information.

2. This chapter was primarily focused on the Web information, which is the major constituent of Big Data today. Come up with a similar discussion about causes, effects, and remedies of the problem of veracity for other sources of Big Data, such as the data arising from the sensors in the Internet of Things or the geospatial data.

33

CHAPTER 3

Approaches to Establishing Veracity of Big Data

Truth is pivotal to many fields of study. Philosophy, logic, and law all delve deep into what constitutes truth. Determining truth is therefore, often multidisciplinary. A number of phenomena from real life and other disciplines are modeled in Computer Science. In this chapter, we shall examine a broad overview of the frameworks and techniques that have the potential to detect the veracity aspects of data. The techniques are broadly categorized into Machine Learning, Change Detection, Optimization techniques to minimize the cost of lies, Natural Language Processing, Formal Methods, Fuzzy Logic, Collaborative Filtering, Similarity Measures, and of course, the promise of Blockchain Technology. We start with Machine Learning, which is often talked about as the "Mortar of Modernization" and is sometimes compared to Electricity that revolutionized the world in the late 19th and early 20th centuries.

© Vishnu Pendyala 2018
V. Pendyala, *Veracity of Big Data*, https://doi.org/10.1007/978-1-4842-3633-8_3

Machine Learning

As we have been seeing, veracity of data is a tough problem – NP-Hard, in the Software parlance. NP stands for Non-deterministic Polynomial time, meaning we cannot find an algorithm for the problem, which runs in time that can be expressed as a polynomial. Even if we find an algorithm to solve an NP-Hard problem, it takes tremendous amount of time, so much so that the solution is just infeasible. There is sufficient proof in the research literature confirming that Misinformation Containment (MC) is NP-Hard. Fortunately, when it comes to the intricate matters of life, math comes to the rescue. Math is the heart of matter. Once expressed in math, the matter dissolves and yields, just like when you touch people's hearts, they dissolve and yield. But before jumping into mathematical abstractions, let us consider how this is done in the real world. How do we learn that something is true in real life? As mentioned earlier, a lot of Computer Science solutions are derived from the real world or inspired by the real-world phenomenon. It is often said that data is like people – interrogate it hard enough and it will tell you whatever you want to hear. In serious crimes, that is what law enforcement officers do to the accused to extract truth. We need to build the right kind of framework to interrogate the Big Data – that is the whole idea of Big Data Analytics.

Note   The problem of veracity of Big Data can be seen as quite analogous to the same problem in law enforcement. Instead of people, we have data and tools in the former.

Establishing truth is the crux of legal battles. How do we tell if someone is telling the truth? We examine the associated details – some features of the scenario. For instance, a lie detector examines features like the pulse rate, blood pressure, breathing rate etc. Similarly, Big Data has some features that the analytics software will examine. Machine Learning algorithms used in the analytics software depend on features extracted from the real world to classify the data. The features are indicative of the class to which the data belongs. Features therefore are crucial to building a model to detect truth. Let us now examine the steps in the process of evaluating the truthfulness.

Note   Like people, data has features that can reveal the truth.

To develop a mathematical model, let us take the instance of a fruit vendor. Assume that we are machines without any prior knowledge or intuition. How do we know that the apples that the vendor is selling are really apples and oranges are indeed oranges and that he is not telling a lie? It is from their features like color, shape, size, texture of the skin, hardness, and smell, having observed a number of apples, oranges, and other fruits before, which are labeled and known as such. The features are not constants – their values are not binary. They vary. We cannot expect any feature like the shape or color with certainty. For instance, we cannot for sure say that oranges are all perfectly spherical or apples are all red. Oranges could be oval as well and apples can come in maroon and other shades of red. The best tool to model such uncertainty is probability. Each feature such as shape and color can be treated as a random variable with a probability distribution. We can also call them predictor variables because they help in prediction or as independent variables like the independent variable 'x' in a coordinate system.

Note   Probability is the primary tool for modeling the uncertain characteristics of data.

We pretty much know the probability distribution of the features of the fruits – we know what sizes and shapes apples come in, by observing several apples from a set of apples. We call this set as the "training set" because we train ourselves in recognizing apples by examining the features of the apples in this set. The variable that is dependent on these features is the category of the fruits, in this case, apples. Given the features of a new fruit, the problem now is to determine its category - whether it is an apple, orange, pear, or something else. We need to build a *model* from the training set, which will help predict the category of the new fruit, given its features. Think of model as a black box, a mathematical abstraction, which takes the new fruit as input and tells whether the fruit is an apple, orange or some other fruit. The black box essentially computes a model of its idea of an apple, an orange and so on, based on the labeled fruits it observes in the training set. We call the set of fruits from which the new fruit is drawn the "test set" because the fruits from that set are used to test the model that we built from the training set. We can capture the above discussion in the following equation:

$$P(Y=1|X)=f(X) \tag{1}$$

where

- $X=\{x_1,x_2,\ldots,x_n\}$, the set of random variables x_i

- x_i are the random variables representing the features of the fruit.

- $P(Y=1|X)$ is the probability of a fruit to be, say, an apple, given its features, X.

Note   Features of the data are represented as Random Variables in a Probability Distribution.

Let us move on to building the model now. It can be easily seen that not all features are equally important. For instance, shape is not as important in determining the category as color or skin texture because both oranges and apples can more or less be spherical. So, not all features carry the same importance or weight. We therefore need to weigh each feature in arriving at the decision as to which fruit it is. The greater the weight, the more important is the feature to our decision making. In the above equation (1), $f(X)$ when expressed in terms of weights, is:

$$f(X) = w_0 + \sum_{i=1}^{n} w_i x_i \qquad (2)$$

Note   Features of the data, like features of all real things are only relative in importance and hence need to be weighed when making decisions.

Before we substitute equation (2) into (1), we need to realize that the Left-Hand Side (L.H.S.) of equation (1) is always between 0 and 1, being a probability. So, we need to come up with ways to make the Right-Hand Side (R.H.S) also always to be between 0 and 1. There are a few ways to do this, one of which is to use the "logistic function," expressed as:

$$logistic(z) = \frac{e^z}{1 + e^z} \qquad (3)$$

The above function is always assured to be positive but less than 1, ideal for our use. Using the logistic function in equation (1) will give us a model, which helps in identifying the true classification of the fruits. A model thus built using the logistic function is called a "logistic regression" model. The equation in (1) now transforms to:

$$P(Y = 1 | X) = \frac{e^{w_0 + \sum w_i x_i}}{1 + e^{w_0 + \sum w_i x_i}} \qquad (4)$$

Note   Weights, sometimes also called parameters of the features, play a key role in determining the truthfulness of data.

The problem of determining the truth now reduces to finding the weights $w_0, w_1, ..., w_i, ..., w_n$ in the above equation (4). It is easy to understand that the goal of the weights is to reduce the error arising from the mismatches between observed and the values estimated using the above equation. The class of fruit resulting in the highest estimated probability in the above equation (4) is the chosen class for the observed fruit. If we estimate a fruit to be an apple using the above equation, we want the observed fruit to indeed be an apple. If it turns out to be an orange, the estimation is in error. We use what is called the Maximum Likelihood to reduce the chance of this error and thereby determine the weights in the above equation, by applying the above equation to the training set. The word "Maximum" should bring the Calculus lesson on "Maxima and Minima" to mind. Indeed, the same methods are used in computing the weights using the Maximum Likelihood principle.

Note   The goal of the selection of the weights is to reduce the chance of error in our determination of the veracity of data.

When we are given a new fruit with given features, we substitute the weights derived from the training set and the features extracted from the new fruit in the above equation (4) to determine if the fruit is indeed what the vendor is trying to sell it as. The above analogy applies to any Big Data set. The process is summarized in Figure 3-1 below. Big Data can be in the form of social media posts like tweets, measurements from sensors in the IoT devices, or other records containing valuable data. Features can be extracted and results annotated (labeled) using manual or automatic processes. It is important to note that the same set of features needs to be used for training the model as well as for testing it. Crowdsourcing tools such as Amazon Mechanical Turk are typically used for manual annotations and feature extraction. Feature extraction is often done automatically by methods not covered in this book. The process of coming up with features, formally called Feature Engineering, is a study by itself. In fact, the efficiency of Machine Learning algorithms depends substantially on the features extracted. For the purposes of the examples in this book, we assume that the features are extracted manually.

Note   An important step in Machine Learning is annotating the training data.

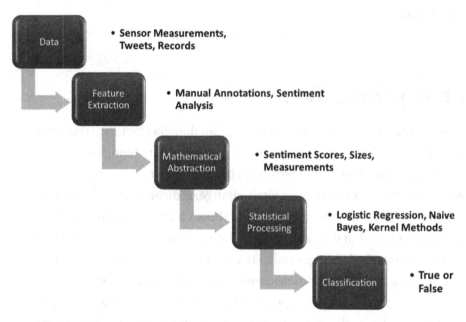

Figure 3-1. *The Machine Learning Process of determining truth of Big Data*

An example of automated feature extraction, say, in case of social media posts like tweets, is by using sentiment analysis, which gives out sentiment scores for each tweet.

Once the features such as the sentiment scores are expressed in math, we no longer have to deal with the original data. We can use statistical techniques such as logistic regression used in the above example, on this mathematical abstraction to arrive at the classification. The entire process is shown in Figure 3-1.

Note   Determining the truthfulness of Big Data using Machine Learning techniques, like solving a math problem, is a step-by-step process.

Change Detection

As we saw in the Machine Learning section, data has features. Features play a key role in verifying truth. In the case of human beings telling a lie, their features such as heart beat, breathing, perspiration, pulse rate, and other stress response vitals change, indicating that the person is lying. Lie detectors essentially detect these changes to convey the truthfulness of what the person said. The same applies to certain classes of data. Unjustified changes indicate possible untruthfulness or fraud. For instance, if the sensor measurements go out of whack for a certain duration or period, there is a possibility of a malicious attack. The chapter on this topic later talks about how malicious attacks on Microblogging websites can also be detected by analyzing the fluctuations in the sentiments of the microblogs posted.

Note   Monitoring changes helps detect lies.

Math helps in this class of techniques as well. Some of the techniques discussed in this book to detect changes include Sequential Probability Ratio Test (SPRT); the CUSUM technique; and the Kalman Filter, which is sometimes considered as the greatest discovery of the 20th century. The intuition behind these change detection techniques is simple. Data often follows a pattern or is more or less constant. For instance, data from the sensors measuring the vitals of a patient under constant observation do not normally change beyond a certain range. The pulse rate, heart rate, blood pressure, arterial oxygen level, body temperature, and respiration rate are often constant or fluctuate slightly.

A good estimate for the next data point in such cases is often the current observed value, indicating status quo. To accommodate fluctuations, the next observed values need to be incorporated into the scheme for coming up with the estimate. So, we start with the estimate as the first value. The next estimate can be the average of the previous estimated value and the current observed value and so on. In the process, we observe the changes in the value, as compared to the estimated value and analyze these changes to determine if there is a possibility of loss of integrity. For instance, if the observed value is greater than the estimated value by a certain threshold value, there is a possibility of abnormality.

Not all changes may be a result of an abnormality, so change detection is a bit more involved than looking for a change that differs from the previous value by a certain threshold. The changes need to be analyzed more deeply for more accurate results. A number of change detection techniques such as Kalman Filter, CUSUM, and SPRT can be used to analyze the changes in more reliable ways. A broad overview of this scheme is illustrated in Figure 3-2. Observed or given data is analyzed for changes using estimation or other statistical techniques. Kalman Filter uses estimation, whereas CUSUM uses cumulative sum to analyze changes. Depending on the analysis, the data may be flagged as false or genuine.

Note   Like in Machine Learning, estimation or anticipating the true value is the key in some change detection techniques.

Figure 3-2. *Overview of Change Detection approach to Veracity of Big Data*

The above simplistic scheme of estimating by using the average of the estimated and observed value misses learning from mistakes. A feedback loop is often needed to improve the accuracy of estimation. Techniques such as the Kalman Filter include a feedback loop to improve the estimation, in addition to other improvements in arriving at the estimate. The next chapter discusses these details. The feedback loop is essentially a measurement update – an update to the estimate based on the observed or measured value. The feedback loop is illustrated in Figure 3-3.

Note   Adjusting the behavior based on the feedback from the performance is a key systems concept that is characteristic of some algorithms like the Kalman Filter, which can be used for change detection.

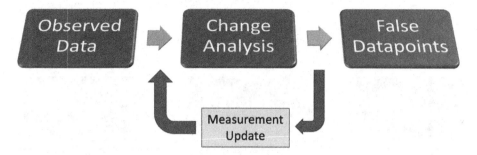

Figure 3-3. *A feedback loop is inserted to improve the estimation*

While techniques like the Kalman Filter help in detecting changes, not all data flagged as false by the algorithm is indeed false. The algorithms are impacted by false positives. Some algorithms like CUSUM perform best on offline data. For these reasons, a single change detection algorithm is sometimes not sufficient to accurately uncover anomalies. An ensemble of one of more algorithms is used for better accuracy in such cases. Each of the algorithms can be fine-tuned for better accuracy by adjusting certain parameters that are ingrained in the algorithm. Just like with Machine Learning, such fine-tuning parameters are predetermined using a "training set" of data. The parameters can then be used on the incoming real-time data.

Note   The Kalman Filter is really an estimation algorithm that can also be used for change detection.

The process of determining the fine-tuning parameters is illustrated in Figure 3-4. The algorithms are successively run on the training data to obtain the parameters. The training data already has the data points annotated as genuine and false. The algorithms are run with a number of possibilities for the parameter values; and the combination, which results in the best match for the annotated training data is retained for subsequent use on the real-time data, also called as "test data."

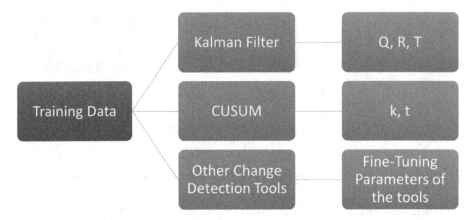

Figure 3-4. *Fine-tuning parameters of the algorithms are obtained on the training data*

An ensemble of the algorithms can then be used to overcome the limitations of one another. For instance, CUSUM works best on offline data but is less impacted by false positives. On the other hand, the Kalman Filter is impacted by false positives but works on real-time data. The two algorithms can be complemented by each other when used as an ensemble. The ensemble approach is illustrated in Figure 3-5. The incoming data is first processed by the Kalman Filter algorithm in real-time. If the algorithm indicates a possible compromise of the integrity of the data, the CUSUM algorithm is invoked on the data that has been received so far. If CUSUM does not indicate that the data has been tampered with, the processing stops, implying that the Kalman Filter algorithm resulted in a false positive. Otherwise, other Change Detection schemes can be further leveraged to assess if the data is indeed impacted by adverse quality.

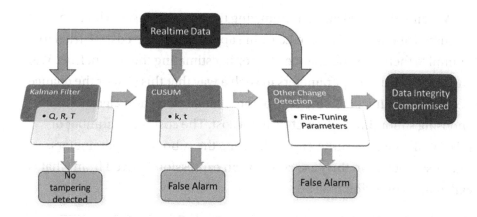

Figure 3-5. *Ensemble approach to determine the Veracity of Big Data using Change Detection Techniques*

Another approach to integrating the algorithms is to implement a simple voting mechanism, where the data is processed by various algorithms independently and data is assumed to be compromised only if the majority of the algorithms indicate tampering. We will examine the Change Detection schemes and ensemble methods in more detail in a later chapter.

Optimization Techniques

In the above discussion on Change Detection techniques, we saw that fine-turning parameters are obtained by doing several trials of the algorithm on training data, which is already annotated as true or false. Iterating over many possible values bluntly is inefficient. There is a need to optimize the way we find the parameters. Optimization techniques come into play wherever there is scope for improvement. The goal of optimization is to minimize inefficiencies or "costs" associated with a way of doing things. Mathematically, cost can be abstracted as a function over a number of variables. Many problems can be solved when expressed as an optimization problem. The veracity problem is no different.

We already talked about minimizing the cost passingly, when we discussed about the Machine Learning approaches. The cost we tried to minimize there was the chance of error in estimating the type of fruit. We chose the weights or parameters in such a way that this cost or the chance of error is minimized. In the case of the Change Detection algorithms, choosing suboptimal parameters has a cost. The cost is the number of misclassifications resulting from choosing the suboptimal parameters. The cost function in this case could be an expression derived from what is called the "confusion matrix."

Note   Lies and compromised data have a cost – the cost of falsity. Optimization techniques help choose algorithms and methods, which minimize this cost by solving sub-problems toward this goal.

A confusion matrix, sometimes called an "error matrix" is a simple two-dimensional table with observed values as rows and estimated values as columns. Looking at the table, we can know how "confused" the algorithm we deployed for determining the truth was. If the algorithm performed well, the prediction (columns) match the observed values (rows). The confusion matrix is an easy visualization of how the algorithm performed. The confusion matrix helps to visualize the true positives, false positives, true negatives, and false negatives. The terms are illustrated in Figure 3-6. False Positives are also called type I errors and False Negatives are called type II errors.

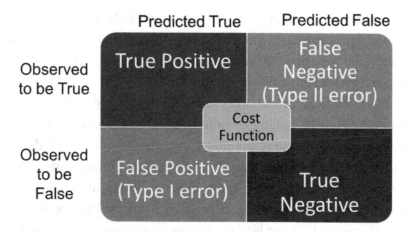

Figure 3-6. *Cost Function can be computed from the various performance metrics*

An example of a False Negative condition or type II error is when a fire alarm does not ring when there is fire. False Positive is when the alarm rings where is no fire. Like in the case of the fire alarm, veracity of data is essentially a classification problem – the data needs to be classified as true or false. Incorrect classification has a cost that varies with the domain. A commonly used component of the cost is the error rate, which is as defined in equation (5) below:

$$\xi = \frac{(FP + FN)}{(TP + TN + FP + FN)} \tag{5}$$

where

$FP = Number\ of\ False\ Positives;$

$FN = Number\ of\ False\ Negatives;$

$TP = Number\ of\ True\ Positives;$

$TN = Number\ of\ True\ Negatives$

For some domains such as health, understandably, false positives or type I errors are many times more expensive than the false negatives in some cases. For instance, if a diagnostic test falsely shows that a patient's blood sugar level is high and is given insulin based on that assumption, it can prove to be fatal. On the other hand, if a test falsely shows that the blood sugar level is not high, when it indeed is, the patient may still tolerate the condition until symptoms worsen and another test is done. In some other cases such as not identifying a malignant tumor early, it can prove to be fatal. A false positive in this case may not be as costly as it may only lead to further examinations, which may conclude that it was a false positive. Alternatively, it can lead to an unnecessary surgical removal of a benign tumor, which also does not have that high cost.

A different kind of example is that of the fire alarm that we considered earlier. In this case, false negatives are more disastrous than false positives. Often, but not always, status quo is less expensive than arriving at incorrect conclusions. Such considerations need to be taken into account, when arriving at the cost function. A simple error rate may not always be suitable. Parameters reflecting the domain characteristics need to be included in the cost function. From the fruit vendor example and the logistic regression model that we talked earlier, when discussing about Machine Learning approach, it can be easily seen that the cost varies with the weights that we choose for the features.

Note   False Positives usually have a higher cost.

For instance, if we weigh the shape of the fruit higher than the texture of its skin, there is a chance that oranges can also be categorized as apples, thereby increasing the cost of the model. The weights should be chosen to minimize the cost. In fact, Machine Learning models choose the parameters or weights that minimize the cost of the model, when applied to the training data. Training data already has the correct classifications or observed values since it is already labeled, so the error rate that we talked above can be computed. A Machine Learning model essentially expresses the cost in terms of the parameters or weights, often represented by θ and tries to find the θ for which the cost is minimum.

Often, there are constraints that need to be satisfied, when trying to find the parameters that minimize the cost. A typical constraint could be that the parameters or weights need to be greater than 0. It is fairly difficult to solve optimization problems, even when the cost function and the constraint functions are "smooth." Smooth functions have the same intuitive meaning of being continuous when plotted, without any jerks. Mathematically, it means that the function has continuous derivatives (slopes) in a given domain. Albeit being difficult to solve, there are efficient methods to solve optimization problems.

Machine Learning models are a specific case of Optimization Frameworks. Optimization techniques span way beyond the problems that Machine Learning models can solve. Optimization is in fact a universal problem. From atoms combining into molecules to optimize energy to attaining peace and prosperity in life, optimization plays a role in most of the universal phenomena. Therefore, it is not always the cost that is minimized in optimization problems. The more general term for the function that is optimized is "objective function." Once we have the objective function and the functions that define any constraints on the parameters, the job of the optimization framework is to minimize or maximize the objective function. If the objective function is cost, we choose the parameters to minimize it.

Note   Depending on the domain, optimization techniques beyond what are used in typical Machine Learning models may need to be used to solve the veracity problem.

Natural Language Processing

A substantial amount of Big Data is in natural languages such as English and Chinese. The veracity problem is particularly rampant in this form of data, contributed by Web 2.0 – the social media. Ironically, the same data can be used to find the veracity of numerical data elsewhere, such as in a tax filing. Recently, the Central Board of Direct Taxes, Government of India awarded a $100 million project to L&T Infotech to detect tax evasion based on social media posts. The company plans to create a "systematic web on a person" based on the information available about the person and his family, on the Web. The "systematic web" will then be used to identify patterns that indicate tax evasion.

Veracity of text data is a tougher problem than that of the numerical data. One way of working with text data is to generate numbers from it, such as by doing a sentiment analysis. Interestingly, text itself has been proven to have certain statistical properties. For instance, text in any language follows Zipf's law, which states that the frequency of a word in a large corpus such as that of a book is inversely proportional to its rank in the frequency table. An interesting corollary of this statistical property is that a substantial portion of any given language comprises a small set of words. As an example, in the English language, just 10 words such as "the" and "be" fill up 25% of the corpus, which is the universal collection of all English text in this world. The statistical distribution of the words in a large corpus therefore follows what is known as the Power law, where one variable varies as the power of another.

Statistics and probability can be used to detect unique styles of language and therefore lies, because often, fraudulent language follows some specific styles. A useful tool for this is the Language Model, which is the probability distribution of sequences of words. Given a sequence of words, $w_1, w_2, \ldots w_i, \ldots, w_n$, a Language Model can give the probability $P(w_1, w_2, \ldots w_i, \ldots, w_n)$ of the words occurring in a sequence. Given the

Language Models of training text data classified as genuine or fraudulent, Machine Learning classification techniques can be used to determine if the Language Model of a given test text data indicates falsity.

Note   Text data demonstrates numerical properties and is surprisingly amenable to numerical and computational processing.

A related tool is the PCFG for Probabilistic Context Free Grammar. Grammar is a set of rules to construct sentences of a language. Context Free Grammar (CFG) is when these language rules can be used irrespective of the context. Context is everything other than the expression itself, which helps us in understanding the expression. CFG ignores the context and only cares about the rules used to generate a language. A PCFG is when the rules have a probability associated with each of them. PCFG is quite efficient in analyzing the syntactic style of a given text. Introducing probability into CFGs gives more flexibility in deriving the language models. The PCFG can be trained using a training set of text, labeled as genuine or fraudulent. For a new set of text, the parser produced by PCFG can be compared with the ones generated from the training set to determine the veracity.

Note   A common theme across several techniques that address the problem of veracity is probability.

One of the characteristics of authenticity is formality. People wear formal clothing such as suits and ties when meeting new people, transacting important business, or giving speeches. Formality has been a sign of trustworthiness even in language. Linguistic formality is closely related to logical or mathematical formality, in terms of the rigor it imposes on the expression. There are a number of NLP tools to measure the formality of a given text segment. The formality score can be used as an indication of the veracity of the text.

Text can be compared for style and structure, when it is represented in suitable format. One such representation is the *Parse Thicket*. A Parse Thicket is a graph that captures the syntactic detail of a given text segment. Parsing is the activity of analyzing a sentence and capturing its syntax in a data structure such as a tree that enables the sentence to be processed computationally. Thicket is the English word for a dense group of trees. A Parse Thicket is therefore a dense group of parse trees for each of the sentences in a text segment. Parse Thickets can be applied in a number of ways to solve real-world problems, one of which is veracity. For instance, plagiarism can be detected by searching for common subgraphs between two text segments. Truthfulness of a given text can be established by comparing its Parse Thicket with that of the text known to be true.

The above are syntactic tools. A good semantic linguistic technique that can be used for determining the veracity of a given text is the textual entailment – analyzing whether the given sentences semantically follow others. When a sentence follows from true sentences, it is more likely to be true. Similarly, a sentence entailed from sentences known to be false can be presumed to be false. There are a number of tools for Recognizing Textual Entailment (RTE). The key here again is the representation of the text that can be computationally processed. We shall examine some of the tools and representations in a later chapter.

Note   Knowledge Representation is the key to computational processing and to verifying the underlying truthfulness of the knowledge.

Formal Methods

Can we impose a syntax for truthful information? The initial answer from many is a clear "no." But as we shall see, in Mathematical Logic, establishing the veracity of a statement is a matter of syntactic processing. Therefore, the question is not completely out of whack and is probably key to future research to establish a truthful world. As we said earlier, math is the heart of matter. Once expressed in math, the matter dissolves and yields. Formal methods help us do precisely the same. They provide us the tool set to represent real-world systems as mathematical entities. When represented as mathematical entities, the systems can be evaluated and their properties verified objectively at a much deeper and subtler level.

Formal methods concisely describe the behavior of a system in ways that are useful for reasoning about them. The systems are specified in a language with standardized syntax and well-defined semantics to support formal inference, which is inferring a statement from a set of other true statements. Machine Learning embraced imprecision and ambiguity. We switch gears here and require that for formal methods like Boolean Logic to work, the knowledge known to be true is expressed clearly and precisely, without any ambiguity or uncertainty. There is also an extension of Boolean Logic, called Fuzzy Logic that can deal with imprecision and ambiguity. Formal Methods can be categorized into Formal Specifications, Proofs, Model Checking, and Abstraction.

Particularly relevant to the problem of veracity is the Logic Specification category of Formal Methods. Logic is the branch of Mathematics that helps us reason about systems, their specifications, and properties. Propositional Calculus, also called as Zeroth-Order-Logic (ZOL), is a basic specification system that allows us to express statements; relation between those statements; and using those, compute the truth of a compound statement. Truth can be checked by truth tables, row by row.

If a logical sentence has n propositional symbols (think of them as English statements) connected together using operators like *or* and *and*, then to check the truthfulness of the sentence, we need 2^n rows to deal with permutations of the two values, *true* and *false*, which each of the n symbols can take. This is clearly intractable, computationally. In fact, this the first problem that has been proven to be *NP – Complete* – NP standing for Non-deterministic Polynomial time, which we discussed earlier as well. Moreover, we can achieve very little with the limited expressivity of the Propositional Calculus system. We therefore consider the next order of logic, the *First – Order – Logic* (FOL) system.

FOL is more expressive than Propositional Calculus. It allows for quantification over variables. The quantification operators used in FOL are for-all, written as $\forall$ and there-exists, $\exists$. These additional operators enable dealing with all or some entities. In FOL, truth can be deduced by using rules of inference in the form of a '*proof*'. A proof is a sequence of statements that can successfully argue the truthfulness of a given statement or lack thereof. An example of a rule used in such proofs is called '*modus ponens*'. The rule simply states that *if p then q and p is true, then q is true.* However intuitive and basic it may sound, the rule is quite powerful and the entire language called *Prolog* is based on this fundamental rule.

A *sound* logic system is one where if a sentence is provable, it must also be true. A sound system will never prove a sentence that is not true. On the other hand, a *complete* system is able to prove every sentence that is true. If a sentence is true in a complete system, it is provable. Inference of truth in a sound and complete FOL has been proven to be intractable, computationally infeasible. Therefore, only a small subset of the FOL is used in reasoning systems. For instance, in Prolog, inferencing is mostly done using just the *modus ponens* rule as pointed out earlier.

Note   Logic, for the most part, is intractable to the computers. The human mind far excels machines in softer aspects like logic.

Knowledge Base (KB) of a system, known to be true, contains facts and rules. For instance, an example of a simple fact in Prolog is

$$female(Mary)$$
$$parent(Mary,Ann)$$
$$female(Coco)$$
$$parent(Coco,Mary)$$

and that of rules is

$$mother(X,Y):-parent(X,Y),female(X)$$
$$grandmother(X,Y):-mother(X,Z),parent(Z,Y)$$

Now if we are asked if Coco is the grandmother of Ann, written as

$$grandmother(Coco,Ann)?$$

By applying the *modus ponens* inference rule and the above two rules in the KB, it can be deduced that the statement is true indeed. As can be seen from the above, deducing truth becomes a matter of syntactic processing, provided the system is specified appropriately. Hence Knowledge Representation or Specification plays a key role in the problem of veracity of Big Data.

Fuzzy Logic

As we saw above, reasoning using sound and complete logic systems, even when the representation is precise and unambiguous, is intractable to computers. Ambiguity, uncertainty, and imprecision make a system much more complex for humans to deal with, than when the system is completely and accurately specified. Surprisingly, machines are able to deal with imprecision far more better than we anticipate, thanks to the many inventions in this space. One of the first sciences invented to deal with imprecision and uncertainty is called Fuzzy Logic. It is more attuned to the way humans think and behave in a non-idealistic world. For instance, if there is a 60% possibility of the first keynote speaker, Raman arriving today and 20% possibility of the second keynote speaker, William arriving today, what do we conclude about the possibility of both Raman and William arriving today? Common sense tells us that there is a 20% possibility of both arriving today, since that is the minimum percent possibility of either.

In Fuzzy Logic, the above commonsense inference is abstracted as the fuzzy *and* operator. Using Fuzzy Logic notation, the above can be written as:

$$\mu_{R \cap W}(today) = \min(\mu_R(today), \mu_W(today))$$

where $\mu_{R \cap W}(today)$ is the *membership function* of today being in the intersection of the set of days that Raman can arrive and the set of days William can arrive. Using similar common sense, we can write

$$\mu_{R \cup W}(today) = \max(\mu_R(today), \mu_W(today))$$

Here, $\mu_{R \cup W}$ is the possibility of Raman or William arriving today. Reasoning proceeds similar to how it is done in traditional logic we covered in the Formal Methods section, but incorporating the membership function calculations as highlighted above. Reasoning with Fuzzy Logic is approximate, so it is called *approximate reasoning*. In fact, reasoning using traditional logic can be considered a border case of approximate reasoning, the border case being when all the membership functions are equal to 1 or 0.

As can be seen from the above description, inferencing truth under uncertain conditions is a matter of possibility. Fuzzy Logic can only infer truth to a certain degree. In the above conference keynote example, it can only infer that there is a 20% possibility of Raman and William arriving today, whereas traditional logic would have inferred either 100% or 0% possibility of both arriving, if at all it had 100% accurate information needed for the prediction. Fuzzy Logic systems can be confirmed to be working correctly by substituting one or zero for the membership function and making sure that the behavior is equivalent to the traditional logic system. Whereas the traditional logic system can only handle just one perfect case when the membership function is 1or 0, Fuzzy Logic provides the tooling needed to deal with those and all the cases in between.

Note   Fuzzy Logic can be considered a superset or an extension of Crisp Logic, also called Boolean Logic.

Information Retrieval Techniques

Truth finding sometimes involves Information Retrieval. When we are unsure about a new statement, our brain retrieves information from the past to compare and contrast that information with the new statement. The science of Information Retrieval provides a number of computational tools to achieve the same job of the brain, using machines. Two such tools are the *Vector Space Model* (*VSM*) and *Collaborative Filtering*. To find the similarities between two documents, we again resort to math, more specifically, to Vector Algebra and of course, to representing information using Vector Algebra principles.

In Vector Space Model (VSM) used for *Text Mining*, each document is represented as a vector in a multidimensional space. Each axis in the multidimensional space represents a word, also called a *term* in the Information Retrieval parlance. If the corpus contains 10,000 terms, there will be 10,000 axes in our multidimensional vector space. To represent a document in this high-dimensional vector space, we need a value across each dimension for the document, like a coordinate in a Cartesian plane.

This value is often the *TF. IDF* of a term. TF stands for Term Frequency and IDF, for Inverse Document Frequency. TF is the number of times the term (same as word) occurs in a given document. IDF indicates how unique the term is. IDF is given by the following expression:

$$IDF_t = \log\left(\frac{N}{df_t}\right)$$

where

> *N is the number of documents in the collection and*
> *df_t is the number of documents in which the term, t appears*

TF.IDF is simply the product of TF and IDF. Today's computing machinery is powerful enough to compute these values in no time, even if the corpus is fairly extensive. Once the TF.IDF is obtained for each term with respect to a given document, it can be plotted as a vector in the high-dimensional space. Once plotted as vectors, the problem of finding similarity of documents reduces to finding the proximity between vectors. There are a number of ways to find the closeness of vectors, the popular among them being the cosine similarity, which measures the angle between two given vectors. Vector Space Model (VSM) is particularly useful to solve veracity problems such as plagiarism and also to establish the trustworthiness of information, by comparing a new document with known sources of truth.

Note   Even in the Information Retrieval approach to the veracity problem, the solution lies in representing the data using mathematical structures and processing them efficiently.

A popularly used technique in Recommender Systems is *Collaborative Filtering* (*CF*). We use this technique in many ways in our day-to-day lives: choosing items from a 'recent returns' shelf in a library instead of going in and searching, listening to the top 10 songs liked by others, hiking on used paths through the woods, citing good research papers in ours, and so on. The common theme in all these examples is that we rely on others' tastes, work, or research for doing similar things as them. Collaborative Filtering (CF) is the act of filtering information and making our choices by relying on others' judgments, thereby indirectly collaborating with them. The classic application of CF traditionally has been to predict whether a user will like an item that she has never judged before, given the judgments about the product from a set of other users.

It can easily be seen how CF can help with the problem of veracity. At a simple level, for instance, the trustworthiness of a vendor can be determined by comparing his prices with those of other vendors. CF can be effectively used for trust propagation. Just like we form opinions and shape our work based on others' judgments, trustworthiness can also be determined based on others' opinions. It must be noted that CF is quite different from voting schemes. As we shall see in a subsequent chapter, CF uses similarity measures such as the cosine similarity that we discussed above.

Blockchain

A new paradigm of impeccable trust that has emerged in recent times is the Blockchain technology, which is the basis for the revolutionary cryptocurrency. Given the investment that has gone into cryptocurrency and the trust it has garnered from all over the world, it is clear that Blockchain is a promising technology that may solve the problem of veracity to a great extent. Facebook recently announced that it will use Blockchain technology to counter fake news. There are a few other companies working on the problem using Blockchain. The very fact that the technology is the basis for the all-important currency that does not need any intermediaries nor government intervention implies the trustworthiness and promise of the technology in solving the veracity problem.

A blockchain is essentially that – a chain of blocks – similar to the linked list data structure, but with some special characteristics. In Blockchain technology, history is ensured to be immutable by using cryptographic hash functions. The *hash chains* that are used have an interesting property that an element in the chain cannot be changed without changing every other element. The blockchain serves as a distributed ledger shared by all participants. Any disagreements that arise in the transactions are resolved by a consensus mechanism. In some ways, Blockchain technology has some ingredients of Wikipedia, which make them trustworthy. However, they are fundamentally different in that Wikipedia's database is centralized, whereas blockchain is based on a Peer-to-Peer (*P2P*) network.

Note   Blockchain technology brought with it a mechanism to build systems and processes that can be completely trusted to perfection.

Summary

Veracity is a problem in many domains and can be addressed in a number of ways. In this chapter, we saw some of the approaches that can be taken to solve the problem. The contents of the chapter are bound to raise the curiosity levels of the reader, with a quite a few questions unanswered. In the subsequent chapters, we shall see more details about each of these approaches and answer questions that this chapter may have raised. We will start with Change Detection techniques in the next chapter.

EXERCISES

1. In this chapter, we talked about the Indian Government awarding a $100 million contract for detecting tax evasion signs from data on Social Media. Search for more such examples of how Big Data is helping with the problem of veracity.

2. There are many more approaches to verifying the integrity of information that have been used in other applications. An example from the Distributed Computing area is the *Byzantine Generals Model*. Describe how it can be used in the context of veracity of Big Data.

3. Examine the feasibility of using Blockchain technology for the different varieties of Big Data.

CHAPTER 4

Change Detection Techniques

Falsity and inaccuracy in Big Data manifests in different forms. In this chapter, we examine one kind of such manifestation and consider ways to identify it. This manifestation is particularly common in the information posted on the Web. It is not uncommon to influence online friends and followers one way or the other through the information posted on the Web. It is a problem when that influence has a hidden agenda, is based on lies, is vicious in intent, and impacts huge populations. This is the manifestation of falsity that the techniques presented in this chapter address. The techniques can also be applied to other scenarios such as when the quality of the data from the sensors in the Internet of Things drops for short periods of time.

Only truth sustains. All else fades into the oblivion. A lie cannot last for long. Truth is pure. It is the foundation of human communication, coexistence, and the civilization itself. Truth can be viewed as the status quo, the norm. Falsehood, then, can be considered as a change, an impurity, a deviation from the expected. Drawing the analogy from electrical engineering, truth is the signal and lies are noise. Lies therefore can be discerned from the truth by using change and noise detection techniques. As a simple example, consider a software program that analyzes school essays. If all the school essays allude to an elephant as a four-legged animal and one of the essays describes the elephant as a bird,

© Vishnu Pendyala 2018

V. Pendyala, *Veracity of Big Data*, https://doi.org/10.1007/978-1-4842-3633-8_4

the latter can be concluded as being false. If the program tabulates the description of the elephant given in different essays, the false claim in the essay that describes an elephant as a bird will stand out and appear as a change or noise.

Change detection is a commonly used ritual in many engineering, biological, and financial processes. The techniques are primarily statistical in nature. We use these techniques on the time series of data points obtained from the quantification process as described in the third chapter on "Approaches to Establishing Veracity of Big Data." In the case of microblogging, the time series comprises the scores obtained from sentiment analysis. Microblogs posted by deploying hacks to influence and sway the public opinion are bound to have out-of-whack sentiment scores, so they can be detected using some of the techniques described in this chapter.

Note   A lie is a change from the norm. Truth is a matter of time. No lie can hide from time. Untruthfulness is a spike, a hype that cannot last for long.

Microblogs are proven to have substantial impact on public opinion. We have seen their influence in the presidential elections in the United States, right from the days when Barack Obama first contested in 2008. Posting false, disruptive information on microblogging websites like Twitter is akin to calling "Fire" in a movie theater. Andrés Sepúlveda realized the role microblogging plays in important social processes like the presidential elections. "When I realized that people believe what the Internet says more than reality, I discovered that I had the power to make people believe almost anything," he said.

Figure 4-1. *Graphical depiction of untruthfulness in the time series of Big Data*

True to his conviction, for almost a decade, people believed in his hacked attacks of injected influence on a microblogging website. The injected influence swayed the outcomes of presidential elections in South American countries for 10 years, and its perpetrator, Andrés Sepúlveda, is currently serving 10 years in prison for the crime. The story is another powerful message on the importance of the veracity of Big Data. If microblogging websites had checks in place to detect such injected influence attacks, the results of the presidential elections that Andrés Sepúlveda swayed may have been different. The checks are not too difficult to implement, as we shall see in this chapter.

Note   An Injected Influence Attack on a microblog website can be defined as the activity of posting deliberate, malicious microblogs, often but not always, using automated means and hacking, in order to impact the opinions of the other users.

Figure 4-1 is one interpretation of falsity in Big Data. The lines are not meant to be straight as the quality of data does vary from time to time. But those fluctuations are minimal compared to when there is a preponderance of dishonesty. To detect changes in the time series of inaccurate and uncertain data points such as those shown in the figure, we need a way to capture the representative state of all the data points until any given time. It is the smallest artifact that completely and recursively summarizes all the previous states of a given system. This representative state should accumulate the relevant aspects of the state until a given time. Monitoring this representative state will help in detecting suspicious changes.

An analogy is a person going into depression. Bad feelings, failures, and various other factors impacting the psyche of the person accumulate over time, eventually resulting in depression. When a person who is normally happy and upbeat is sad, we are able to detect the change and try to prevent it. Happiness is the norm or the "signal," and sadness is the change or "noise" that can be detected. The state of the human mind of the person is representative of all the happenings over the timeline. Detection of the progression of mental depression is possible if we were to monitor the state of the human mind. When the mind is sad and depressed beyond a point, an undesirable change has occurred and corrective action should be necessitated.

Note   For Change Detection to work, we need a way to model the representative state of the time series of the Big Data values.

The techniques described in this chapter facilitate the process described in the above analogy, as applied to data. The techniques are statistical, so can work only on numbers. Quantification is therefore essential if the data such as microblogs are not numerical. Even if numerical, it is important that the value of the data itself is accurate. So, valuation, as we introduced in the first chapter, also plays an important role. The change detection techniques presented in this chapter also assume certain statistical properties of the underlying data.

A commonplace statistical property that is assumed is the "Gaussian Assumption" that the data fits into a normal probability distribution. The mean, mode, and median of a Gaussian distribution are all equal. The assumption is not too constraining because most of the natural and social sciences data is anyway expected to converge to the normal distribution, particularly if drawn from substantially large samples as found in Big Data. Interested readers can refer to the central limit theorem for more details. Classical statistics plays an important role in quite a few valuation schemes as well. Big Data theory therefore requires quite some background in theoretical probability and statistics.

An advantage of the techniques presented in this chapter is that they are weakly correlated and not too sensitive to the probability distribution of the underlying data. We do not have to exactingly ensure that the data is normally distributed to apply these techniques. Moreover, studies have shown that Big Data, such as those from the microblogs, have certain statistical properties, and opinion mining is feasible by performing sentiment analysis on the microblogs. For instance, a surprising finding by the American linguist, George Kingsley Zipf (1902–1950) is that the frequency of a word of the English language, or for that matter any natural language, in any large corpus is inversely proportional to the word's rank in the descending order of frequencies.

As an example, the word, "the" is the most commonly used word in the Corpus of Contemporary American English (COCA), followed by the word "be." According to Zipf's finding, "the" should occur approximately twice as many times as "be" in COCA, which surprisingly is true. When the corpus follows this Zipf's law it is said to be approximated by a Zipfian probability distribution, which is a kind of power law distribution. Power law distributions are characterized by a long tail in the plot and only a few values dominate the distribution. True to this characteristic, in the Brown corpus of American English text, it is found that only 135 words account for half of the corpus and many words are used only a few times, giving the distribution a long tail for the many words that are less used. Studies have proven that the microblog data confirms to Zipf's law, exhibiting the statistical characteristics of a Zipfian distribution. Assured by these findings, we now proceed to examine some of the techniques used for change detection.

> **Note**   Surprisingly, there is order to the seemingly chaotic data
> around us, particularly when the data is big as in Big Data.

Now that we are convinced Big Data has certain statistical properties, we proceed with ways to monitor the data to identify falsity. A brute force, coarser, and simple approach to detect changes is to monitor the values for any out-of-whack patterns and conclude that the data has been manipulated, based on the observations. But by doing so, we are ignoring the uncertainty component involved in the problem. What if the out-of-whack patterns were indeed genuine and not a result of any manipulation? For instance, a person becoming irritable and deeply sad may not necessarily mean he is getting into depression. Occasional bouts of mood swings are not uncommon. It is therefore not enough to just monitor the time series values or in this example, moods. We need to monitor the representative state of the entire time series till then. Hence the need for robust algorithms.

We are not sure if a particular change really implies falsity. We need to involve probability to deal with this uncertainty. As we shall see in the following sections, using statistics and probability to monitor the data stream also makes it faster to detect changes. The algorithm is much more sensitive to changes. In fact, we may be able to detect an undesirable change as soon as the manipulation or falsity starts by using probability, as contrasted to observing the out-of-whack patterns for a while to come to the same conclusion. We shall first start with a test that has been quite popular since its invention in 1945.

Sequential Probability Ratio Test (SPRT)

SPRT was invented by Abraham Wald and published in his paper, "Sequential Tests of Statistical Hypotheses" in June 1945. He proved that the test is optimal in the Speed versus Accuracy tradeoff (SAT) and that it requires very few data points to detect changes, as long as the data points are independent and identically distributed, a common assumption abbreviated as i.i.d.

As in statistical hypothesis testing, SPRT formulates two hypotheses and evaluates the time series data against the backdrop of these two hypotheses:

H_0: The quality of data is not compromised and is acceptably true;

H_1: The data is false, trigger an alarm;

H_0 is called the null hypothesis – the status quo or the norm and H_1 is called the alternate hypothesis indicating a change.

In the Sequential Probability Ratio Test, for every data point, we compute the log-likelihood ratio (LLR). The LLR is the logarithm of the likelihood ratio. There are four components in the LLR:

a. Likelihood of getting the data point from the valuation process, given H_1 is true and the data has been manipulated

b. Likelihood of getting the data point from the valuation process, given H_0 is true and the data is not compromised

c. Ratio of the above two, (a) and (b)

d. Logarithm of the ratio in (c)

In equation form,

$$LLR = \log \frac{p(x_k \mid H_1)}{p(x_k \mid H_0)}$$

where x_k is the data value at the kth point in time

(4.1)

The likelihood in the numerator and the denominator depends on the distribution of the underlying data. For instance, if the underlying data confirms to a normal distribution, the Probability Density Function (PDF) is given by

$$p(x|\mu,\sigma) = \frac{1}{\sqrt{2\pi}\sigma}e^{\frac{-(x-\mu)^2}{2\sigma^2}} \tag{4.2}$$

where μ is the mean of the distribution and σ is the standard deviation, which can be easily inferred for a given population. For each data value, the LLR is accumulated in the cumulative LLR, abbreviated as $CLLR_i$ as expressed in equation (4-3) below

$$CLLR_i = CLLR_{i-1} + LLR_i \tag{4.3}$$

We now know why the technique is named SPRT - the process is Sequential (S) and uses Probability (P) Ratios (R) to Test (T) for changes. The cumulative LLR (CLLR) is the representative state of the data collected so far. Monitoring the cumulative LLR can help us detect changes. We use the CLLR to decide one of the following three outcomes:

a.   accept H_0 and stop monitoring the data, concluding that the data quality is intact

b.   accept H_1 and stop monitoring the data, concluding that falsity has occurred

c.   defer the decision and continue monitoring

When each of the above decisions is triggered depends on two common, intuitive types of errors:

a.   type I: accepting H_1 when H_0 is true, resulting in a false positive

b.   type II: accepting H_0 when H_1 is true, resulting in a false negative

We need to decide what level of type I and type II errors can be tolerated by the system, beforehand. Let α denote the acceptable probability of encountering a type I error and β, the acceptable probability of encountering a type II error, determined beforehand. It can be shown that the decision-making boundaries are as follows:

$$A = log \frac{\beta}{(1-\alpha)} \ and \ B = log \frac{(1-\beta)}{\alpha} \tag{4.4}$$

Here are the decisions based on the above boundaries, also depicted in Figure 4-2:

a. if CLLR < A, accept H_o and stop

b. if CLLR > B, accept H_1, conclude that there is falsity and stop

c. if CLLR is in between the two values A and B, defer the decision and keep monitoring

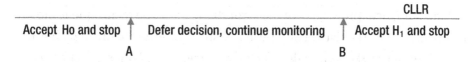

Figure 4-2. *Decision Boundaries in SPRT using CLLR*

Note   Sequential Probability Ratio Test is just that: a test of the accumulated ratio of likelihoods computed sequentially.

As we can see, adding probability into the solution makes it slightly complicated because the monitoring now depends on the distribution of the underlying data and the use of math in a much more involved manner. We need a simpler technique to get the math of exponentials such as in equation (4.2) out of our way from the algorithm. As we shall see in the section below, CUSUM addresses this complexity and makes the algorithm much simpler.

It must be noted that SPRT determines which of the two hypotheses is generating the data and not fundamentally meant to detect changes. It can be modified to detect changes by introducing a "threshold value." When the LLR exceeds this predetermined threshold, we stop monitoring and conclude that a change has been detected. The resulting technique is called CUSUM and discussed in the next section.

The CUSUM Technique

CUSUM is short for Cumulative Sum. It is a simple but clever, intuitive technique invented by E. S. Page in 1954. The technique often finds place in stock portfolio monitoring in finance and process control in engineering to detect shifts in the values being monitored to focus on products that are likely to run into problems. Though the mathematical derivation assumes certain statistical properties of the underlying data, the technique is mostly insensitive to the statistical and probabilistic properties of the data. It is the fastest technique known to detect changes, even small ones. It performs much better without having to compute moving averages involving exponentials such as in equation (4.2), like in SPRT.

Though drastically simple to implement as compared to SPRT, CUSUM, in fact, is a form of SPRT. Instead of two boundaries A and B as in SPRT, we just have one stopping point that occurs when the representative state exceeds a certain threshold. The representative state here is the accumulated residual at each data point. The residual can be thought of simply as the difference between the data value and a target value. The target value is, quite intuitively, the mean of the data values until then. The algorithm therefore is simple arithmetic without having to deal with PDFs and exponential moving averages.

CUSUM is both simple and complex at the same time. We just saw that the application of CUSUM is simple arithmetic. We shall take an intermediary approach, giving just enough details to appreciate the math behind the simplicity. Let us start by giving the equation for the representative state, r_k in the CUSUM algorithm:

$$r_k = \max\left(r_{k-1} + \log \frac{p(x_k|H_1)}{p(x_k|H_0)}, 0 \right)$$

where

    x_k is the data value and

    r_k the respresentative state at the kth point in time

The log ratio in the above equation can be immediately recognized as the LLR that we used in SPRT. The hypotheses too are the same that we used before. The improvement in CUSUM is that we are now accumulating the LLRs and have a lower bound for the accumulated value. When the distribution of the data is Gaussian or normal, which is often the case, we can substitute the following PDF into the equation:

$$p(x_k|H_1) = \frac{1}{\sqrt{2\pi\sigma}} e^{\frac{-(x_k-\mu_1)^2}{2\sigma^2}} \quad and \quad p(x_k|H_0) = \frac{1}{\sqrt{2\pi\sigma}} e^{\frac{-(x_k-\mu_0)^2}{2\sigma^2}}$$

The mean values, μ_0 *and* μ_1 correspond to the two hypotheses respectively. The equation for the representative state, r_k now simplifies to:

$$r_k = \max\left(r_{k-1} + \left(x_k - \mu_0 - \frac{\mu_1 - \mu_0}{2} \right), 0 \right)$$

and further to:

$$r_k = \max\left(r_{k-1} + \left(x_k - \mu_0 - \frac{\rho}{2} \right), 0 \right)$$

where 'ρ' is a constant that determines the sensitivity of the CUSUM algorithm. The smaller the value of ρ, the more sensitive the algorithm is to the change. The above equation is simple arithmetic that we described earlier. We essentially monitor the changing mean of the data values and accumulate the difference or residual of the new data value over the mean. We now need to know when to stop monitoring the data stream and decide that the data has been compromised. For this, we introduce a threshold τ. When the representative state, r_k exceeds the threshold τ, we stop monitoring. The value of ρ when this occurs is called as the stopping time. It must also be pointed out that even though we derived the simplified equation by substituting the Gaussian PDF, it has been proven that the simplified equation works well even when the distribution is not Gaussian.

Note   All complexities in the world are meant to make life simple. The CUSUM derivation is an example.

Often, we do not want to stop as soon as falsity is detected. We want to continue monitoring and detect all occurrences of the mendacity. Instead of determining the stopping time, we want to detect all the intervals when the data is compromised. CUSUM can be easily extended to do this. The previous "stopping time" now becomes "start time" of the suspicious interval. The end time of this interval is when the representative state is at its peak. Going back to the analogy of a person going into depression, we conclude an onset of depression when the mental state of the person starts deteriorating or exceeds a threshold. We conclude that the person is getting out of depression soon after his mental state is at its worst and starts improving. The same applies to the extended CUSUM algorithm to detect all suspicious intervals.

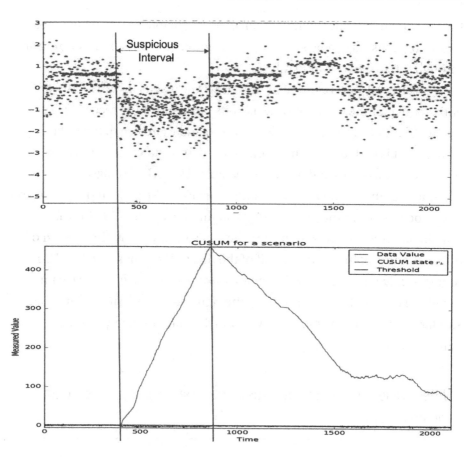

Figure 4-3. *Extended CUSUM to detect suspicious intervals; the dots in the upper plot represent data values*

Figure 4-3 is an illustration of the process that was just discussed. The data values in this example vary between -5 and +5. The upper half of the figure shows the values plotted as discrete points. There are around two thousand data values. The x-axis represents time. It shows the two thousand indexes in time, representing the order in which the data was collected. As can be seen from the first half of the figure, even if possible, it is hard to tell if the data has been compromised just by looking at the data values. Once we plot the CUSUM representative state, r_k in the graph below, shown in lighter shade (green), the suspicious interval becomes almost obvious, as can be seen from the figure. The curve is ascending all through the suspicious interval.

The interval starts as soon as the CUSUM exceeds the threshold and ends soon after the CUSUM value is at its peak and starts receding. The threshold is determined to be close to zero and therefore to the data values, so it is mixed up with the line plot of the data values in slightly a darker shade (blue) and is not very visible. The CUSUM algorithm accumulates the residual of each data value from the expected average and as can be seen from the figure, can go all the way close to 500 for values that do not exceed 5 in magnitude. There is a difference in the pattern of points in the plot in the upper half of the figure, but unless the CUSUM representative state is plotted as in the lower half of the figure, we cannot really be sure. Quite a few points on the right side also form a somewhat similar pattern, but the plot below shows they are not really suspicious, which is indeed the case.

Note   Unless evaluated by a method – a statistical process, data can be deceptive.

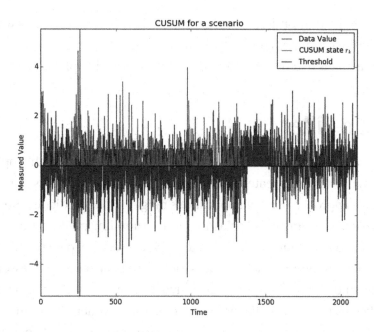

Figure 4-4. *Extended CUSUM when the spurious values occur intermittently. The figure in color is in the source code file and in the ebook.*

In Figure 4-3, all the spurious data occurs at once in a large interval. However, there are cases when the suspicious values occur once every now and then. Figure 4-4 shows the CUSUM state plotted in a lighter shade (green) when the false data values appear intermittently. The suspicious intervals contain just one false value in this case and the same logic still applies. The CUSUM technique is not as effective in this case because the accumulation of the residual is impacted by genuine values coming in between. The maximum accumulated residual in Figure 4-3 was close to 500, whereas in Figure 4-4, it is just 5. CUSUM algorithm is obviously less sensitive in this case.

This is where Kalman Filter fares better. It estimates the next value recursively based on the previous states, so the more genuine values preceding the spurious values, the better the estimate of the next value.

The greater the preponderance of genuine values in the previous states, the better the estimate. The next section discusses this important technique that is extensively used in many real-world applications.

Kalman Filter

M. S. Grewal et al., authors of one of the most popular books on the topic, call Kalman Filter the "greatest discovery of the 20th century." It is a linear state-space approach used in applications ranging from predicting the uncontrolled trajectory of celestial bodies to controlling manufacturing processes and aeronautics. Simply stated, given a series of data points, a Kalman Filter computes the best guess for the next data point. It is essentially an estimator of the next state of a linear dynamic system, when the signal is muddled by Gaussian white noise. As mentioned earlier, Gaussian Assumption is quite common in many stochastic processes and holds in most cases, at least approximately. Since it predicts the signal, a Kalman Filter can automatically be seen as filtering out the noise. Noise is the change that is detected by the Kalman Filter, so in that sense, we are using it as a Change Detection technique here. Understandably, it is extensively used in electrical, telecommunication engineering, and signal processing.

As mentioned at the beginning of the chapter, lies are essentially noise in a civilization that predominantly thrives on truthfulness. Falsity in Big Data can be viewed as statistical noise and detected the same way as noise in a signal processing system is filtered. That is where the Kalman Filter comes in handy. The complex math behind this powerful technique can be understood by taking an intuitive example such as the price of a stock. Say we expected the price of a stock to be $\hat{x}_k$ today, but we observed it to be z_k in reality. Given these two values, our best estimate for tomorrow is the average of these two values:

$$\hat{x}_{k+1} = 0.5z_k + 0.5\hat{x}_k$$

But the rumours and influences in the stock market make it not so ideal, and we may have to weigh them more compared to our own expectations. Let us call this weight w_k. Then the above equation transforms to:

$$\hat{x}_{k+1} = w_k \cdot z_k + (1 - w_k) \cdot \hat{x}_k$$

This is essentially the state equation of the Kalman Filter in the linear case. There is an observed component and an expected component in the right-hand side of the equation. We make a calculated guess for the expected value and update it based on the real-world observation or measurement. The proportionality term or weight, w_k, in the Kalman Filter parlance is called the "Kalman Gain, K_k" in the honor of the inventor, Rudolf Kalman, an MIT graduate. The weight w_k is representative of the noise – rumors and other influences in the stock market.

As we can understand, the weight w_k or the Kalman Gain, K_k cannot be estimated simply by intuition and needs some mathematical rigor as well. It is estimated using another set of forward recursive equations involving one other state variance variable, P_k, and two noise variance constants Q and R. Variance is a commonly used statistical metric to indicate how much the variables deviate from their mean values. Q is the noise variance of the estimated values and R is the noise variance of the observed values. In reality, Q and R can vary with state or observation, but for simplicity, in most cases, they are assumed to be constant. As before, z_k is the observed value. Incorporating all this discussion, the system of equations for the Kalman Filter in the linear case of a time series of data values is given below.

$$\hat{x}_k^- = \hat{x}_{k-1}$$

$$P^-_k = P_{k-1} + Q$$

$$K_k = \frac{P^-_k}{P^-_k + R}$$

$$\hat{x}_k = K_k \cdot z_k + (1 - K_k) \cdot \hat{x}_k^-$$

$$P_k = (1 - K_k) \cdot P^-_k$$

In the general form, the terms in the above equation are all matrices and P_k, Q, and R are covariance matrices. But for simplicity and for application to a single dimension time series of data values, we can assume them to be scalars and variances. Though their values depend on the domain, for simplicity, the initial values of the initial estimated mean $\hat{x}_0$ is set to zero and apriori state variance P_0 can be set to one or determined by intuition. The other variables in the above equation can then be computed recursively. Optimal values of the assumed constants Q and R can be obtained by applying the algorithm to what is known in the Machine Learning parlance as the "training data." In the training data or ground truth, we already know which values are genuine and which are false. So, we calibrate the constants to achieve a high accuracy rate using this sample of known data.

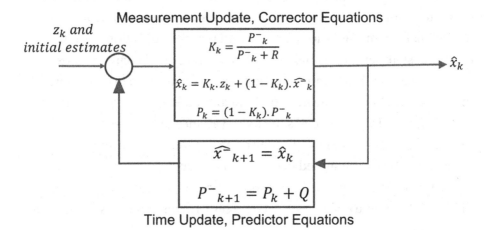

Figure 4-5. *Kalman Filter interpreted as a Feedback Control System*

A Kalman Filter is optimal because it can be proven that it minimizes the mean squared error of the estimated values. Mean Squared Error (MSE) is a common metric used to determine the accuracy of prediction or estimation algorithms. It must be noted that the Kalman Filter equations exist in many forms, all of which can be algebraically shown to be equivalent to each other. The equations essentially implement a Feedback Control System. Feedback from the previous estimates by way of the observed values is incorporated into the next estimate. In that sense, the first two equations can be categorized as 'time update' and the next three as 'measurement update' equations. The latter is to incorporate the feedback and the former to project the current representative state into the future. Therefore, the time update ones can also be called as predictor equations, while the latter as corrector equations. Figure 4-5 illustrates these concepts.

The equations all have intuitive interpretations and are based on the theory of probability. For instance, as the variance R of the inaccuracy in the observed values tends to zero, the Kalman gain K_k tends to one indicating that the observed values z_k can be increasingly trusted. In this case, it also means that the estimated component is weighed lesser and lesser. In the case when the initial state variance P^-_k approaches zero, Kalman gain K_k approaches zero as well, indicating that there is more reliance on the estimated values than the observed values. A probabilistic or Bayesian interpretation of a Kalman Filter is that the recursive equations essentially propagate the conditional probability distribution of the signal in the data stream, given the observed values.

It must also be noted that Q and R converge to steady state values with time. These steady state values can be obtained using the training data, beforehand. R is based on measurements, so it is more easily obtained than Q, which is based on the estimation process. Accuracy of the Kalman Filtering process depends on how well the noise variances Q and R are estimated.

Note   Kalman Filter recursively computes the best estimate for the next data value using a set of mathematical equations, incorporating all information that is available until that time. Both the previously estimated and observed values are weighed into the new estimate.

Figure 4-6 shows how a Kalman Filter helps separate the spurious values from the genuine ones. The data are sentiment scores from the microblogs injected with negative sentiment. The estimated value calculated recursively using the equations listed in the previous paragraphs is shown as a blue line. As can be seen, it is more or less straight and close to the value zero. Since we cannot always expect the real data values to exactly match the estimated values, we allow a margin of tolerance, a fixed offset, shown in the plot as a red line below the blue one. All the data values below the red tolerance line are determined as spurious by the Kalman Filter algorithm. Since the data are sentiment scores from microblogs injected with negative sentiment, we consider only the data below the tolerance line. In other cases, we could have a line above and a line below to delimit the tolerance interval. Data values outside of the interval are considered spurious. In this case of sentiment scores, the spurious values are marked with a red '+' and the genuine values are shown as a '*' in a darker green shade above. The margin of tolerance, which is a fixed offset from the expected values, is again calibrated to maximize the accuracy obtained by trying the algorithm on the training data.

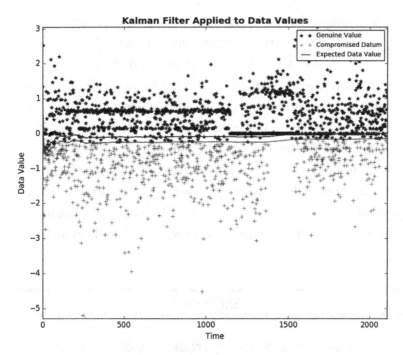

Figure 4-6. *Kalman Filter applied to the same data as in Figure 4-4*

One prerequisite for the Kalman Filter to work correctly is that the signal should be much stronger than the noise, which is typically the case. In the example in Figure 4-6, there are approximately four times the number of green darker shade stars than the red crosses. Most of the values are positive and genuine, so the negative values below a certain margin of tolerance are determined to be spurious, which is mostly correct in this case.

Summary

Untruthfulness is a change from the norm. It can be considered noise in an otherwise truthful valuation of Big Data. We can therefore apply change detection techniques to establish the veracity of Big Data. In this chapter, we examined three interesting and important change detection techniques. SPRT showed how a simple ratio of the likelihoods can prove useful in the detection. CUSUM furthered and simplified this idea to use simple arithmetic of accumulating the residuals to identify spurious intervals of unprecedented changes in the data streams. The discussion on the Kalman Filter showed how noise or falsity that is contrary to the predictions can be filtered out from the signal or the normal data. We learned that training data helps in calibrating important parameters of the algorithm or model. In the next chapter, we will extend the idea of using training data to learn the model itself. We will examine a number of Machine Learning algorithms in the process and apply them to the veracity domain.

EXERCISES

1. What are some of the ways that changes are detected in the real world? Can any of those apply to the problem of Veracity of Big Data? Think of any other real-world processes that can help with the problem.

2. Research how the Kalman Filter is used in various applications. Are you convinced that the Kalman Filter is the "greatest discovery of the twentieth century"?

3. Derive the equations for the Kalman Filter given in this chapter from the state-space representation in vector form, given in any book on the topic.

CHAPTER 5

Machine Learning Algorithms

In the last chapter, we saw techniques that apply at a finer granular level of data, working with individual data values. In this chapter, we consider data in bulk, at a much more macrolevel, to extract features, patterns, and their relative weights. Interestingly Machine Learning is a double-edged sword. It can be used to create fake content and also to detect fake content. A good example is Google's "DeepDream" project. The project started out to build image recognition software. But now, the software is noted more for generating imaginary drawings, as if from hallucinating machines. In this chapter, we will focus on Machine Learning algorithms that can help in detecting fraud and fake information.

Machine Learning is inspired by the concept of human learning. How does a child learn, for instance, to recognize a cat? Her parents or other elders show her many samples of dogs, cats, and other pets and repeatedly identify each pet to the child. In other words, the parents give the child plenty of labeled data for training, also called *training data*. The child starts to identify a number of *features* of these pets – the shape of the face, the texture of the skin, the length of the trunk, the size of the ears, the color of the eyes, and so on. The more number of features the child identifies and the more number of labeled samples she sees, the better the child learns to recognize a cat.

© Vishnu Pendyala 2018
V. Pendyala, *Veracity of Big Data*, https://doi.org/10.1007/978-1-4842-3633-8_5

This learning process of determining the features, their relative importance, and drawing conclusions based on them is not limited to cats and dogs. Almost everything, including distinguishing truth from lies is learned by experience using similar methods. Elders have the practice of "kidding" with children. Soon, with experience, children learn to know when the elders are "kidding" and when they are telling the truth. Though the problem is about determining the truth, deterministic methods are not used for the purpose. Truth is not defined deterministically, providing the child with a number of rules to describe a cat or to decide when the elders are kidding. That process is intractable or *NP-hard* in software terms. Instead, the child learns empirically from a number of samples. Let us look at this process in more detail.

Note   Defining what constitutes truth exactly is intractable. Truth, like faith, is best learned experientially than defined deterministically.

When the child sees a new four-legged animal and there are no elders nearby to identify the animal, the child's learning about the pets is put to *test* with this new data. It sets the thought process in motion, in one of many ways. One way is the *frequentist* approach – the more the frequency of seeing a cat with a round face, green eyes, small ears, short trunk, etc., the more the child is likely to conclude that the new animal with a round face, green eyes, small ears, short trunk, etc., is a cat. This method is formalized in the *naive bayes* approach, an example of the frequentist method.

We can also set forth a *decision tree* – does the animal have a round face? If so, are the eyes green, ears small, etc. The more "intelligently" we frame these questions, the faster we will be able to reach a conclusion. The goal of "intelligence" in this case is to choose the questions that result in the widest separation between the two categories. For instance, if we choose a question about the length of the trunk before checking for the shape of the face, we are bound to generate a decision tree that is suboptimal, because there could be dogs with trunks as short as those of cats. The question about the length of the trunk, therefore does not separate the categories as much as the shape of the face. The latter question therefore should supersede the former question.

Note   Empirical methods such as the Machine Learning algorithms tend to incorporate intelligence to quickly arrive at the best answers, rather than to cover all possibilities by brute force.

A commonly used method of getting to the truth is onion peeling – in layers. We start with the information inputs we currently have and start interrogating in layers. The more complex the problem is, the more the number of layers of questions. The answers to all these questions need to be connected to arrive at the conclusions. This process in the brain of collecting and connecting answers in layers of *neurons* can be simulated on machines using *neural networks*. Neural networks take inputs and, using a series of layers, compute the outcome of classification as true or false.

The eventual goal of all the learning processes or *algorithms* that we consider in this chapter is to partition or *classify* the solution space. The better and clearer the separation, the more effective is the algorithm. While having labeled samples to learn from is good, there are often occasions, when such data is not available. In such cases, learning proceeds *unsupervised* – there are no elders to *supervise* the learning process by showing samples and labeling those samples. An example is the child learning to pick all Barbie dolls from the toy chest with a number of different toys. The child learns to *cluster* her favorite Barbie dolls, without anyone having to tell her how to do it.

Note   Learning can progress supervised or unsupervised. For veracity purposes, the eventual goal of both ways is to divide the problem space into two classes – truthful and not truthful.

Clustering is again based on features. Given data and information that is not entirely true, the *unsupervised learning methods* group the data into true and false buckets, based on the features of the data or information. In this chapter, we shall examine both *supervised* learning methods such as *neural networks* and *unsupervised* learning methods such as $k-means$ to solve the data veracity problem. But first, let us consider a suitable example to illustrate the concepts.

The Microblogging Example

Microblogging websites such as Twitter have evolved as powerful media that can cause substantial impact. In the second chapter, we have seen how Twitter was used to pump up the stocks and dump them for an illegal profit, and lies posted on it caused companies to lose up to 28% of their market capitalization. In Chapter 4, we saw how Andrés Sepúlveda used Twitter to even change the outcome of presidential elections. There are many such examples where fraud on a microblogging website has caused substantial damage. We therefore take the example of Twitter to illustrate how Machine Learning can be used to detect fake tweets.

The problem is illustrated in the plot in Figure 5-1 below. Given a number of tweets, the need is to come up with a *Classifier*, which appears as the curved line that separates the true tweets, shown as stars in the figure from the false tweets, shown as rings. As can be seen, we still use math to express the problem, to make it amenable. The figure is a Cartesian plane with tweets pointed as points. At first glance, the figure raises three questions:

 a.   How can we plot the tweets in the space shown in the figure?

 b.   What should the axes represent in such a space?

 c.   How can we come up with the Classifier?

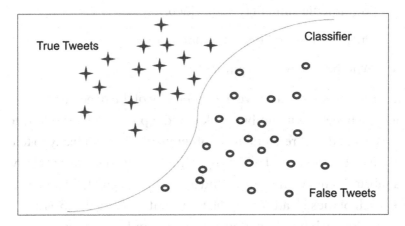

Figure 5-1. *Illustration of the example problem: Separating true from the false tweets*

In Chapters 3 and 4, we learned the ideas of quantification, extracting features, and using mathematical abstraction. Those ideas come in handy when answering the above questions.

Microblogs, such as tweets, have features. Some of the features could be the following:

 •   Number of retweets, replies, likes

 •   Number of hashtags in the tweet

 •   Hashtag used is listed as (worldwide) trending

- Presence of URL and its page rank, authentic entities like NIMH, WHO, NCI

- Presence of reference to another user

- Formalism in the tone of the tweet (no abbreviations like 'd', 'n', etc.)

- Use of affirmative, definitive, absolute words rather than "may be"

- Multimedia content (photo, video, audio)

- Presence of numbers, quotation marks

- Whether the tweet is pinned

These features are indicative of the veracity of the tweet. For instance, if a tweet has a number of likes and replies, it is more likely to be true. A tweet that is referencing another user in it is not likely to tell a lie because the referenced user's reputation is also at stake. Similarly, tweets substantiated with pictures and audio have a greater chance to be true. Each of these features can be represented along the axes and their values for each tweet, plotted, giving us a multidimensional space with all the tweets plotted in the space. For those familiar with Vector Algebra, the multidimensional space can be thought of a *vector* space and the points as *vectors*. Figure 5-1 is a two-dimensional projection of this multidimensional space and the classifier is really a *hypersurface* in the multidimensional space. Since it is hard to visualize a hypersurface when the space has more than three dimensions, orthogonal projections such as shown in Figure 5-1 really help.

The hypersurface divides the multidimensional space into two. In an n-dimensional space, the hypersurface has (n-1) dimensions. For instance, in a two-dimensional space shown in Figure 5-1, the hypersurface is a one-dimensional surface, which is a line (curved or straight) If the space is three-dimensional, the hypersurface is a two-dimensional plane or surface. The equation of a hypersurface is linear. For the two-dimensional case, we know the equation of a straight line has the form

$$y = mx + c$$

Expressing it more generally, we can write the equation as

$$\theta_0 + \theta_1 x_1 + \theta_2 x_2 = 0$$

$$where\ y\ is\ x_2,\ x\ is\ x_1,\ m\ is\ -\frac{\theta_1}{\theta_2}\ and\ c\ is\ -\frac{\theta_0}{\theta_2}$$

Extending the above equation to n-dimensional space, the equation of the hypersurface can be written as

$$\theta_0 + \theta_1 x_1 + \theta_2 x_2 + \ldots + \theta_n x_n = 0$$
$$or\ in\ short,\ \Sigma \theta_i x_i = 0$$

(5.0)

We'll keep encountering (5.0) in different forms subsequently. The above equation separates the multidimensional space into two spaces given by the equations below.

$$\Sigma \theta_i x_i > 0 \ldots \ldots \ldots \ldots \ldots \ldots \ldots \ldots (one\ side\ of\ the\ hypersurface)$$
$$\Sigma \theta_i x_i < 0 \ldots \ldots \ldots \ldots \ldots \ldots \ldots \ldots (second\ side\ of\ the\ hypersurface)$$

Given the features of a new tweet, $X = \{x_1, \ldots, x_i, \ldots, x_n\}$, and after computing the parameters, $\Theta = \{\theta_0, \theta_1, \theta_2, \ldots, \theta_i, \ldots, \theta_n\}$ from the training set (we shall see later, how), we can easily determine on which side a new data point is going to fall based on the above inequalities. We determine the weights or parameters from the training dataset and use them on the test dataset to classify a new data point, just like the child does, to recognize a cat. Of course, the algorithms should make sure that learning is not too dependent on the training dataset - a situation called *overfitting*. We do not want the child to recognize only a certain subset of cats that are very similar to the ones she saw before. We want her to learn to recognize a cat in general.

> **Note**   Machine Learning algorithms can be thought of as treating data as a set of points in a multidimensional space, with each dimension representing a feature of the data.

A few other relevant features pertaining to the user, that help with establishing the veracity of the tweets are listed below.

- Number of followers, lists user is on, tweets over a period, photos and videos in the profile and in tweets

- Social Status: Is the user a celebrity?

- Presence of titles like "dr," "prof," "governor," "mayor," in the handle, verified account

- Trustworthy characteristics of user's profile: authentic name, photo, profession, schooling, associations, date of joining, domain names in the email id (like stanford. edu) etc.

- PageRank of the URL in his profile

- Number of list categories the user is featured in

- Number of posts retweeted, favorited, replied

- Number of references to the user in others' tweets (@user in tweets)

- Number of @user references in this user's tweets; quality of those users

- Number of URLs in his posts (knowledge monger) over a period of time

- Number of followers weighted by their followers and so on – kind of a page rank

Note   The higher the credibility of the user, the more likely are her tweets to be true.

The features are quite self-explanatory and can be easily seen as indicating the truthfulness of the tweets. Using the features, we now have a multidimensional space in which tweets are plotted. Most of this chapter is devoted to arriving at the *classifier*, the line (hypersurface, really) separating the true tweets and the false tweets. But before that, let us take a brief look at an important aspect of the Machine Learning algorithm – collecting the training dataset.

Collecting the Ground Truth

The more samples a child sees, the more he learns to get to the truth. So, the more the labeled dataset we have for training, the better we learn about determining the truth. The training set acts as the *ground truth* in the process of learning. If it is the Twitter dataset, the size of the training set typically is several thousands of tweets. Microblogging websites such as Twitter provide streaming APIs, using which posts, events, and other data are automatically pushed, avoiding expensive polling by the clients. One can also use REST APIs to collect static data from these websites. A few static datasets are also made available by authors of papers, who may have used them for their published projects.

After collecting the dataset for training, the next step is to categorize or *label* the data. Without labeling, unless we use unsupervised or semi-supervised learning algorithms, the dataset cannot be of help in training. Labeling the data requires someone to go through each of the samples and applying the labels after careful examination. This is typically done via crowdsourcing – the process for getting tasks done by a number of people by leveraging the Internet. If there are, say 300,000 samples, we can probably crowdsource 3,000 people to work on 100 samples each. The monetary expense for this depends on the tasks and the people accepting the tasks.

In our microblogging example, if we consider only health-related tweets, which have a high negative impact if not entirely true, this process can be prohibitively expensive. Labeling health-related tweets as true or false requires substantial medical expertise, which we know is scarce and costly. Even assuming that the medical professionals such as nurses in the United States are paid around $50 per hour and can label 30 tweets an hour, 300,000 samples will require 10,000 hours and a whopping half-a-million US dollars!

Note   Collecting the training data can be a substantially expensive operation.

The above example clearly indicates a pressing need to automate the ground truth collection process. One way to do it is to collect the tweets, which we know for sure are true and another set of tweets, which we know for sure are false. For the latter case, for instance, a number of people tweet based on certain established myths. One such myth is that "microwaving kills nutrients." A number of real-time tweets are surprisingly based on this myth proven to be false. We can find tweets like "Bet y'all ain't kno when u microwave food it subtracts the nutrients" and "The oven and stove work just as well. Heat from the stove and dry heat from the oven kill nutrients too, but not as much as the microwave" or 'The fast pressure of life & "convenience" food. Microwave meals cooking that kills off all nutrients' getting likes, retweets, and replies.

Note   Myths and misunderstood truths are still a major source of falsity in public discourse.

False tweets can be collected by searching for such myth-based posts. It can be done by simply searching for phrases from popular myths such as "microwave nutrients." That gives us one class of the dataset. For the true tweets, we similarly search for tweets with patterns known for sure to be true. For instance, tweets with quotation marks citing established medical journals are usually true. Similarly, tweets that bust the myths are known to be true for sure. Such tweets that the author found on Twitter include, "Often microwave your veggies? By doing so, you could be preserving vital nutrients" and "Random potato fact: the best way to preserve the nutrients in a baked potato is to microwave it. Actual science was done #EatWellForLess." Once we collect all such automatically *labeled* tweets, we may need to hand-check a few of them to weed out any false positives.

With some creative thinking, we can thus automate the collection of ground truth in quite a few cases. It must be noted that tweets are only taken as an example to discuss the algorithms. The techniques apply to other forms of data as well – just that the features are different and depend on the data domain. Now that we have our training set, we move on to the next step – that of finding the classifier. There are many Machine Learning algorithms and their variations, which can accomplish this task. We discuss only a few of them to get an idea of what is involved.

Note   Creative thinking is at the heart of automation, including that of collecting the ground truth.

Logistic Regression

We briefly touched upon Logistic Regression in Chapter 3, so let us start with this algorithm and consider more details. In our Twitter example, tweets have a number of features that can be represented on the axes of the multidimensional space as the independent variables. Let us call this vector of features X.

$$X = (x_0, x_1, x_2, \ldots, x_n)$$ (5.1)

We know that not all of these features are equally important. For instance, we saw that tweets that were based on myths are also getting likes, replies, and retweets. The number of likes, replies, and retweets, clearly is not high in importance when it comes to determining the truthfulness – it *weighs* less, compared to, say, the feature if the tweet is pinned. A user does not normally pin a tweet known to be false. It is apparent that the features need to be weighed in to get a score for the truthfulness of the tweet. We compute the score s, as follows:

$$s = \Sigma_{i=0}^{n} w_i x_i$$ (5.2)

where w_i is the weight assigned to the ith feature, x_i

It can be seen that (5.2) is similar to (5.0). In fact, both are identical and weights are indeed the parameters that need to be estimated.

But what we really need is the estimated probability that the tweet is true and the above score s is not guaranteed to be between 0 and 1, which a probability should be. We use a special function called the logistic function to transform this score into a probability. We say that the parametric statistical model adopted for our solution is "logistic regression." The logistic function is defined as

$$logistic(z) = \frac{e^z}{1 + e^z}$$ (5.3)

This function is plotted in Figure 5-2. As can be seen, its value varies from 0 to 1. More specifically, it can be seen from Figure 5-2 that

$$logistic(-\infty) = 0$$
$$logistic(0) = 0.5$$
$$logistic(\infty) = 1$$
$$logistic(-z) = 1 - logistic(z)$$

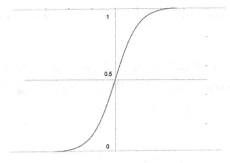

Figure 5-2. *The plot of a logistic function*

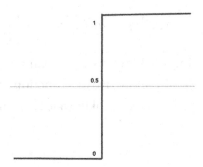

Figure 5-3. *The Step Function*

Another observation from Figure 5-2 is that the plot of a logistic function has the shape of the letter *S*. The class of functions, whose plots have the shape of *S* are commonly referred to as *Sigmoid Functions*.

Figure 5-3 shows a step function. The logistic function plot is quite similar in shape to the step function, with a few distinct and useful characteristics. It is smooth, continuous, and monotonic. Many real-world phenomena fit this model, where values vary continuously rather than discretely as in the case of a step function. The logistic function is also related to the odds ratio. As we know, the word, *odds* refers to the likelihood that an event will happen. If p is the probability that the event will occur, the odds are given by $p/(1-p)$. The inverse of the logistic function, called the logit function is essentially the logarithm of these odds.

Note   The logistic function is closely associated with probability and is quite widely used in Probability, Statistics, and Machine Learning.

$$logit(p) = \log\left(\frac{p}{(1-p)}\right) \tag{5.4}$$

$$logit^{-1}(p) = logistic(p) = \frac{e^p}{1+e^p} \tag{5.5}$$

As can be seen, the logistic function is well-suited for our purpose in more than one way. We therefore use this function to express our truthfulness score as a probability by taking a logistic of the score as follows:

$$P(tweet = TRUE \mid X) = \frac{e^{\sum_{i=0}^{n} w_i x_i}}{\left(1 + e^{\sum_{i=0}^{n} w_i x_i}\right)} \tag{5.6}$$

where X is the set of features, x_i

We just substituted the expression for the score (5.2) in the logistic function (5.3) in place of z. Now the Machine Learning problem reduces to finding the weights w_i given x_i for the tweets in the training data and their respective probabilities of either 1 or 0. Once we know the weights from the training data, we can use the same weights to determine $P(tweet = TRUE \mid X)$ for any new tweet, using(5.6). The operation is illustrated in Figure 5-4.

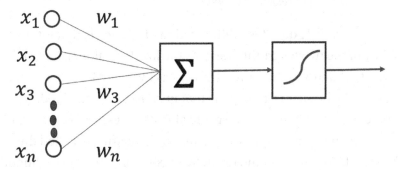

Figure 5-4. *Pictorial representation of equation (5.6) in terms of the features, weights, the sum of their product, and finally substituted in the Sigmoid function*

An observation worth noting is the following rearrangement of (5.6) in (5.7), reiterating the fact that the inverse of the logistic function is the log odds.

$$From\ (5.6),\ P(tweet = TRUE \mid X) * \left(1 + e^{\sum_{i=0}^{n} w_i x_i}\right) = e^{\sum_{i=0}^{n} w_i x_i}$$

$$therefore,\ P(tweet = TRUE \mid X) = e^{\sum_{i=0}^{n} w_i x_i}\left(1 - P(tweet = TRUE \mid X)\right)$$

$$and\ \frac{P(tweet = TRUE \mid X)}{\left(1 - P(tweet = TRUE \mid X)\right)} = e^{\sum_{i=0}^{n} w_i x_i} \tag{5.7}$$

We can readily recognize the L.H.S of (5.7) as the *odds*.

The problem of finding the weights, also called *parameter estimation*, can be solved by using what is known as the Maximum Likelihood Estimate (MLE), which aims to maximize the likelihood with respect to the weights. MLE is a basic concept of statistics, used to estimate parameters. Mathematically, we find weights w_i such that the product of all probabilities given by the following expression is maximum.

$$\Pi P \left(\begin{array}{l} \textit{tweet's truthfulness, given its features} \\ \textit{and their respective weights} \end{array} \right) = \Pi P(Y \,|\, X, W)$$

More explicitly, Π denotes multiplication of the terms that follow P, the probability of the tweet's truthfulness; Y, the labeled value (true or false) of the tweet, '$|$' denotes the word "given"; X is the set of features for that tweet; and W is the set of the weights that need to be determined.

For the purposes of maximizing, the above expression is easier and equivalent, even when we take the expression's logarithm, called the log-likelihood. We get the same result by maximizing the log-likelihood. The same weights, which maximize the logarithm of the expression in (5.7) also maximize the function itself. Taking the logarithm (log) is useful because the log function changes the above product into a sum, noting that

$$\log(ab) = \log(a) + \log(b)$$

Solving the maximization is simplified further by the fact that Y can take only two values, 1 or 0. Optimization techniques can be used to determine the weights, which maximize (5.7). Using the weights we obtain from the training data, for a new tweet, we can compute

$$P(\textit{tweet} = \textit{true} \,|\, \textit{its set of features } X)$$

Note   The complex problem of Machine Learning is often reduced to the seemingly simple problem of finding the weights or parameters of the chosen model: in this case, Logistic Regression.

Naïve Bayes Classifier

All through our discussion of the veracity problem so far, probability played a key role. Central to the study of probability is the Bayes theorem. This section will explore how a direct application of the Bayes theorem can help solve the veracity problem.

In Logistic Regression, we presumed that the logistic function helps us differentiate between true and false tweets. We directly estimated the score for the truthfulness of a tweet, also an indication of what is called a *conditional probability,* as a linear function of features in equation (5.2). We did not learn what constitutes a truthful tweet and what constitutes a false tweet. We were just concerned about how to differentiate between the two. This is called the *discriminative approach* to Machine Learning. This is a much more practical approach, in case we just need to distinguish between two classes. For instance, if we have to distinguish between two spoken languages, say English and Chinese, it is much more practical to go by the sounds of the words than to learn each language. However, in certain cases, we do learn the languages to determine to which language a given sentence belongs. This is called the *generative approach.* The Naïve Bayes technique falls in this category of generative algorithms.

On one hand, the generative approach is so called because we are able to generate data points ourselves. For instance, if we learn Chinese and English to distinguish between the two, we are also able to generate our own sentences – not just distinguish between the two. On the other hand, discriminative approaches can only draw a distinction between two classes but will not be able to generate new data points. In the probability parlance, generative models learn the prior probabilities and conditional probability distributions from the training data, whereas in discriminative models such as Logistic Regression, we did not attempt to compute such probabilities. "Learning" a class such as the language English in the world of probability is essentially to compute the probability distribution functions pertaining to the class.

Note   Machine Learning algorithms can be classified as discriminative or generative, the latter having the ability to generate data points.

Learning the language example is only to illustrate the distinction between generative and discriminative approaches and should not be taken literally. In fact, often, generative approaches produce less accurate results than the discriminative ones, whereas in the learning the language example, intuitively, generative methods should produce better results, since the languages are learned. So, the analogy was only indicative - not exacting.

In our tweets veracity problem, with the feature set X, there are a few probabilities involved:

$$p(X|Y) = probability\ of\ the\ tweets\ having\ features\ X,$$
$$given\ that\ the\ tweet\ is\ true\ (or\ false)$$

$$p(X,Y)\ the\ joint\ probability\ of\ a\ tweet$$
$$with\ the\ given\ features\ is\ true\ (or\ false)$$

$$p(Y) = \Sigma_x p(X,Y)\ is\ the\ probability\ of\ occurrence\ of$$
$$true\ (or\ false)\ tweets\ in\ the\ dataset$$

> $p(Y|X) = $ *the conditional probability that a tweet*
> *with features X is true* (*or false*)

$p(X) = $ *probability of occurrence of a tweet with features X*

The veracity problem is to compute $p(Y|X)$. The probability of a true (or false) tweet to have features X is $p(X|Y)$. In the frequentist approach, $p(Y)$ tells us how frequently a true (or false) tweet appears in the given dataset. $p(X)$ is really inconsequential because it is the same for true tweets and false tweets but is still used to compute our goal, $p(Y|X)$, as we see below.

Readers conversant with probability will remember the fundamental theorem given by the 18th-century English statistician, Thomas Bayes, given by the equation (5.8) below.

$$\text{Bayes theorem: } p(X,Y) = p(X|Y)*p(Y) = p(Y|X)*p(X) \qquad (5.8)$$

Our goal of finding the probability of a tweet being true, given its features can be achieved using the above Bayes theorem (5.8) as expressed in (5.9) below.

$$p(Y|X) = \frac{p(X|Y)*p(Y)}{p(X)} \qquad (5.9)$$

Most algorithms are based on certain assumptions. One of the underlying assumptions of the Naïve Bayes classifier is that the features are all independent of each other. That means, for instance, having more number of retweets has no relation to having more number of likes. This may not necessarily be completely true in the real world, but it is a fair approximation in most cases. The independence assumption leads to the equation in (5.10) below.

$$p(X|Y) = p(x_1|Y)*p(x_2|Y)*p(x_3|Y)*...*p(x_n|Y) \qquad (5.10)$$

Example: p(tweet having multimedia content, is pinned, references another user| tweet is labeled as true)
    = p(tweet has multimedia content | tweet is labeled as true)*
    p(tweet is pinned | tweet is labeled as true)*
    p(tweet references another user |tweet is labeled as true)
    Similarly,

$$p(X) = p(x_1)^* p(x_2)^* ...^* p(x_n)$$
(5.11)

Note   Past frequencies, counted as the number of occurrences, convert to probabilities and probabilities help predict the future.

The probabilities on the right-hand side of (5.10) can be computed from the training data by noting the frequencies of the respective occurrences. For instance, the training data can tell us how many tweets are pinned that are labeled as true. The Conditional probability then, is simply the ratio of the number of these occurrences to the total number of true tweets in the dataset. Using (5.10), we can therefore compute $p(X|Y)$. Similarly, $p(X)$ can be computed using (5.11). $p(x_i)$ is simply the ratio of the number of tweets with a particular feature divided by the total number of tweets. We multiply these probabilities for each feature together, to get $p(X)$.

The only term on the right-hand side of (5.9) that needs to be computed before computing our goal on the left-hand side, is $p(Y)$. This is simply the probability of finding a random tweet to be true (or false) in a given dataset. It is analogous to computing the probability of finding a red ball in a bag containing m red and n green balls. For the training data, $p(Y)$ is simply the ratio of the number of tweets labeled as true (or false) and the total number of tweets in the training dataset. Now that we have all the terms on the right-hand side, the left-hand side can be easily computed, which gives the desired probability of a new tweet to be true, given its features.

Note   The complex problem of Machine Learning can sometimes be solved by a simple application of the Bayes theorem, with a naïve assumption that the features are conditionally independent of each other, given the label.

Support Vector Machine

A relatively new Machine Learning algorithm, popularized in the 1990s for the task of classification, in our case, into true and false data points, is called *Support Vector Machine* (*SVM*). The idea behind SVM is to draw the classifier in Figure 5-1 in such a way so as to maximize the margin of the separated training data. Margin is the distance between the data points on either sides that are closest to the classifier. Let us get into more details using our Twitter example.

As discussed in the beginning of the chapter, let us assume that the training data has been plotted as shown in Figure 5-5. We have true tweets plotted on the left and the false tweets to the right.

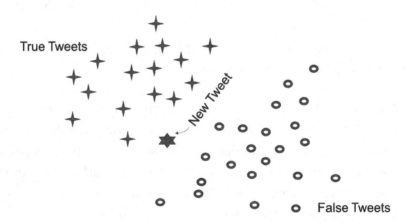

Figure 5-5. *A new tweet falls in between the True and False tweets*

Now we want to find the truthfulness of a new tweet that falls between the true and false tweets in the plot, as shown in Figure 5-5. There are multiple ways to come up with a classifier that separates the true and false tweets in the training set. Let us first limit ourselves to linear classifiers. Figure 5-6 illustrates some such possible linear classifiers. Depending on which classifier we choose, the new tweet can be labeled as true or false. For instance, choosing either classifier 2 or classifier 5 in the figure will cause the new tweet to be labeled false, while choosing any other classifier will cause the new tweet to be labeled true.

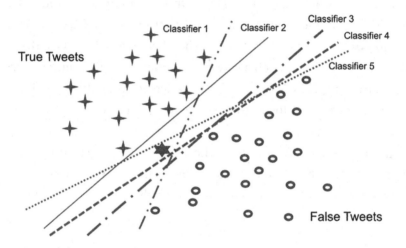

Figure 5-6. *The new tweet can be classified differently by the various possible classifiers that we can choose from*

We clearly need a more decisive way of choosing the classifier. Ideally, the classifier should separate out the true tweets from the false tweets as much as possible. This problem of separating out the tweets as much as possible can be expressed in math, much more clearly as follows. We want to find a classifier such that the distance between the closest true tweets and the closest false tweets is as wide as possible. For that, we identify the points, more punctiliously called vectors, that are on the boundaries separating the two classes. In Figure 5-7, these points or vectors are shown as *Support Vectors*. We want to identify a hypersurface that is at a maximum distance from these vectors. Figure 5-8 shows such a hypersurface, also called the *Maximum Margin Classifier*. The hypersurface is so called because it maximizes the margin between the boundaries of the two classes.

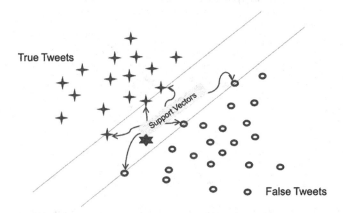

Figure 5-7. *Vectors that are on the boundaries*

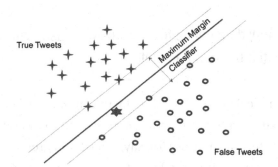

Figure 5-8. *The Hypersurface that is at a maximum distance from both boundaries*

Figure 5-9 illustrates why the vectors on the boundary are called Support Vectors – they can be visualized as supporting the classifying hypersurface. If these vectors are moved even slightly, the hypersurface will have to move with them. A related point to note is that the hypersurface depends only on these points, and none of the other data points in the training set, unless of course those points cross the boundaries to become support vectors.

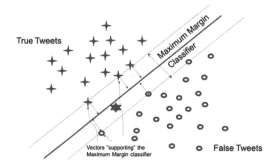

Figure 5-9. *The vectors on the boundaries can be seen as "supporting" the Maximum Margin Classifier hypersurface*

Note   Just a few data points from the past dataset can help predict the class of all the future data points.

The mathematical equation of the Maximum Margin Classifier is still of the form of (5.0). Just that the parameters are found differently, using some different criteria than those for, say, Logistic Regression. However, it must be noted that we simplified the discussion by considering only linear classifiers. That may not always be the case. For instance, in Figure 5-10, the decision boundary is nowhere close to being linear. The hypersurface equation in (5.0) is no longer applicable. In such cases, we use what are known as *kernel functions* to model the classifier. The math needed for finding the Maximum Margin uses Quadratic Programming, which is beyond the scope of the book. However, software implementing SVMs can be mastered even without knowing the intricacies of Quadratic Programming.

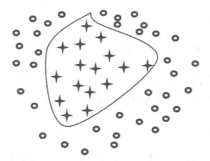

Figure 5-10. *Class boundaries may not always be linear, as can be seen in this case*

Artificial Neural Networks

Our brain functions by connecting a number of neurons and transmitting signals between the neurons using these connections. Learning progresses gradually through the neurons, their connections, and the signals passing through them. Let us see more closely how learning is implemented in our algorithms. Figure 5-11 illustrates what happens in our Machine Learning algorithms such as Logistic Regression. We extract a number of features from the training data set and use them as inputs, x_i. The inputs are factored in using weights for each input, x_i. The weights and the inputs are combined using a function, such as the logistic function, to produce the output classification as true or false.

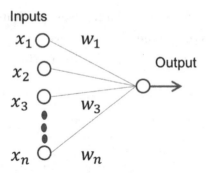

Inputs

Figure 5-11. *A number of input nodes connect to the output node, which does some processing and classifies a new datapoint as true or false*

Using the human brain analogy, each of the inputs can be considered a neuron, more correctly, an *artificial neuron* that takes an input. Each of the input neuron connects to another neuron, which does some processing and produces the output. The node producing the output is called the output neuron. This is simple learning. It is intuitive that the output can be made more accurate by introducing more neurons between the input and output layer, more connections, and more processing. Their function is intermediary – they neither gather inputs from external sources, nor output to external sources. Hence, they are *hidden* to the external world. Accordingly, they can be called *hidden nodes*. Figure 5-12 shows one such arrangement of the *artificial neurons*. The *hidden layer* has three nodes in this case. Going back to our Twitter example, let us assume the three hidden nodes answer the following three questions: a) How reputed is the user posting the tweet? b) Did the tweet attract a lot of attention? c) How authentic is the content of the tweet? These three answers are obtained from the nodes in the input layers, which are really representing features. To answer the first question, the weights corresponding to the user-related features are expected to be high in magnitude, while the other weights are low.

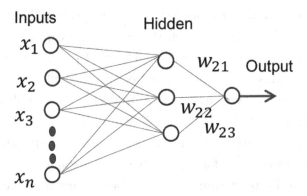

Figure 5-12. *Introducing the "Hidden Layer" of neurons for more detailed processing*

Note   Artificial Neural Networks demonstrate how close Machine Learning is to the biological process of human learning.

Similarly, to answer the second and third questions, the weights for the user-related features are assigned a value low in magnitude. These weights are not shown in the figure to avoid cluttering. For the second question, the weights for the features related to the number of retweets, likes, and replies are assigned values high in magnitude. The third question requires the weights to be high for the features about the content such as if it contains multimedia or references to other users, journals, etc. The values of these three hidden nodes are combined with their own weights, shown in the figure as w_{21}, w_{22}, *and* w_{23} to arrive at the final classification as either true or false. The networks such as those shown in Figure 5-12 are called *Artificial Neural Networks* (*ANN*). They can contain any number of hidden layers. The number of hidden layers and output nodes is determined empirically. For our case, the output node of course will be only one, since the classification is binary – either true or false.

The more the number of hidden layers, the deeper is the learning, leading to *deep learning*. The preceding discussion may raise the question as to how Logistic Regression is different from Neural Network processing. Indeed, Figure 5-11 is a representation of Logistic Regression, similar to Figure 5-4. In that sense, Logistic Regression can be considered as the simplest Artificial Neural Network (ANN). The processing done at the hidden and output nodes of an ANN is similar to what is done in Logistic regression. The processing at these nodes involves transforming the weighted features using a sigmoid function, usually the logistic function itself. The neurons in a multilayer ANN can be considered as logistic regression units themselves. Neural networks are therefore useful when the feature set is huge and there are a number of intermediary problems to be solved before the final classification.

Note   Artificial Neural Networks are many times more powerful than some other Machine Learning algorithms, such as Logistic Regression. Hence the increased interest in Deep Learning, which is based on Neural Networks.

K-Means Clustering

So far, we have been assuming that we are given a labeled dataset that can be used for training the model to compute the weights. However, as discussed initially, getting the training dataset is not always easy, and we need to classify the data solely based on unlabeled data. This is one of the tasks of *unsupervised learning*. One of the widely used unsupervised learning methods is the $K-means\ Clustering$, where items are automatically grouped into different clusters. The resulting dataset is partitioned into K-clusters or subsets, with data in each subset sharing common characteristics. For our veracity problem, K=2, since our classification is binary and the two clusters are true and false.

Note the use of the word *cluster* instead of *class*. Classes are human defined, whereas clusters are determined automatically without any human involvement. But for our microblogging problem, we want the clusters to represent human-defined classes of either true tweets or false tweets. This can be achieved by choosing the features appropriately such that all true tweets have similar characteristics, and all false tweets have similar characteristics different from the true tweets. When plotted, all true tweets are close to each other, and false tweets are also close to each other but apart from the true tweets. As we saw in Figure 5-1, this is indeed the case. The K-means algorithm finds the two groups without resorting to coming up with a classifier as in Figure 5-1.

In the absence of any labeled training data, Figure 5-1 looks like Figure 5-13 –points or vectors in the space representing tweets all look the same and belong to the same class to start with. The problem now is to separate them into two classes: True tweets and False tweets, each of which are bunched together by the nature of their features. Like we saw in Support Vector Machines (SVM), the goal is for the two sets identified to be as apart as possible. As with SVMs, the process of separation needs to use the notion of distance. There are several ways to measure the distance between two points or vectors or data points. Some of them include Canberra Metric, Correlation Coefficient, Cosine Similarity, and Manhattan Distance. To keep our discussion intuitive and also aligned with what is popularly used, we limit ourselves to the Euclidean distance, which is the common geometric notion of distance between two points.

Figure 5-13. *Unsupervised Learning: All tweets are unlabeled and undistinguishable to start with*

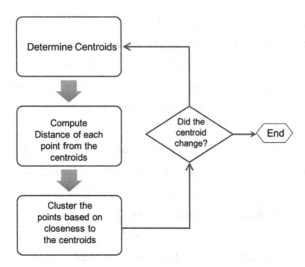

Figure 5-14. *The K-Means Algorithm Flowchart*

K-Means is an iterative algorithm. Clustering is based on the notion of the centroid, which can be visualized as the center of the cluster. Mathematically,

$$\mu(C) = \frac{\Sigma \vec{x}_i}{|C|}$$

where $\mu(C)$ is the centroid (mean) of Cluster C
$|C|$ is the number of points in Cluster C and
x_i are the values of the features of the points in Cluster C

In our case, K=2, so to start with, two points are randomly chosen as centroids of the two clusters. For each other point, two distances are computed: a) the distance from centroid-1, and b) the distance from centroid-2. If the point is closer to centroid-1, it is grouped with cluster-1, otherwise cluster-2. With this new grouping, the center is bound to be changed. New centroids are computed for the new clusters. The process

116

of computing new centroids and regrouping the points continues until all the points remain with the clusters they were assigned to in the previous iteration and centroids do not change.

The K-means algorithm is guaranteed to converge, the proof of which is beyond the scope of this book. It falls under the class of what are known as coordinate descent algorithms. We are assured that the centroids do not change after some iterations, when the algorithm stops. The process is illustrated in Figure 5-14. The clusters around these centroids are the groups of true tweets and false tweets.

Note   Classification into true and false data points can progress unsupervised.

Summary

Learning to distinguish truth from lies is an interesting Machine Learning problem. We saw a few algorithms that help achieve this. The algorithms are either supervised by providing labeled training data or are unsupervised. Independent of this, the algorithms can be either generative or discriminative, the former being able to synthesize data points after learning the probability distributions. An essential element of the discriminative approaches is parameter estimation. The model parameters need to be estimated in such a way that they optimize the accuracy of the algorithms. In the next chapter, we shall change course and examine how logic and formal methods help us in the veracity domain.

EXERCISES

1. Sigmoid functions such as the logistic function played a major role throughout the discussion in this chapter. What are some of the other functions that can possibly take the place of the Sigmoid function and help with Machine Learning?

2. This chapter covered a few Machine Learning algorithms that can be used to solve the veracity problem. Explore other algorithms such as Random Forests and Linear Discriminant Analysis for their suitability to the veracity problem.

3. K-means algorithm assumes that the mean is representative of the cluster. In the real world, however, most often, the "most" vociferous, "most" influential, "most" wealthy, or some other "most" _____ person gets elected to represent the people of a constituency. Examine the performance, in terms of accuracy and the usefulness, in terms of applications, of a clustering algorithm based on the "mode" instead of the "mean."

4. Implement an Artificial Neural Network to solve the veracity of the tweets problem discussed in this chapter.

5. Compare the performance and suitableness of the various algorithms discussed in this chapter to the problem of veracity of microblogs.

CHAPTER 6

Formal Methods

Can we impose a syntax for truthful information? When the author asked this question to the audiences in some of the conferences where he delivered keynotes, the answer predominantly was "no." Indeed, common sense says that truth relates more to semantics than syntax. Handling semantics, particularly of natural languages, is far more complex than dealing with syntax, which makes truth finding a difficult task. Math comes to the rescue, again, now in a different form. The math in this chapter strikes at the crux of the problem – representation of information. Representation of information plays a pivotal role in the algorithms, which process the information. The more precise and formal the representation, the more amenable is the information to automated processing.

Representation makes a difference in real life as well. When thoughts are represented in aggressive, belligerent, confronting language, the problem escalates. The same thoughts, when expressed in helpful, truthful, and supportive sentences can actually solve the problem. Those familiar with Data Structures in Computer Science can readily relate to how the representation of data as linked lists, stacks, or hashes makes a huge difference in solving the problem. In this chapter, we will examine the logic notation to represent data and address the problem of veracity using this representation. A good example of using logic in Big Data is with the *triple stores* used in *semantic web*. Triple store uses *Resource Description Framework* (RDF) for data representation. The techniques presented in this chapter require a similar shift in the representation of Big Data.

© Vishnu Pendyala 2018
V. Pendyala, *Veracity of Big Data*, https://doi.org/10.1007/978-1-4842-3633-8_6

The previous approaches were post-fact – we examined something that had already occurred in the past and determined its truthfulness. The approaches so far could not directly be used to prevent misinformation. They only helped in misinformation containment. If we can impose a syntax to truthful information, we can possibly prevent misinformation from even entering into the system. After all, truth was the only form of information to start with. It is only after the thoughts, interactions, and deeds started getting corrupted, that people invented language constructs to express misinformation.

If the representation can be restricted again to a strict set of *sound* constructs, it may be possible to build totally sound systems, without a place for imprecision and falsity. These ideas are not as far-fetched as they seem. Blockchain technology, discussed in the last chapter, is an example that comes close to the idea of building a completely sound, trustable system. Soon, it may be possible to build entirely sound systems, by exploiting the representation constructs. The representation should of course be backed by well-defined semantics that facilitate deducing truth or the lack of it.

Note   Knowledge and information representation has always played a major role in evolving systems and will continue to do so.

Formal methods are mathematical techniques used for the rigorous representation of computing systems to help with their design and *verification*. The methods are formal in the sense that the specification and operations are precise and conforming to rules. Data about the behavior and properties of systems is captured in concise syntax that aids reasoning about the systems. Mathematical Logic is one of the constituents of Formal Methods. As we shall see, logic is all about syntax and axioms and rules to do operations with the syntax. Verification, reasoning, or proving the truth becomes a matter of strategic application of axioms and rules, transforming the information to *entail* truth or lack of it. Data, when represented using logic constructs, helps the veracity problem by transforming it into a logical reasoning problem.

Note Formal methods provide powerful techniques to translate general descriptions about a system into a precise specification language using standardized syntax and reason with it in increasingly automatic ways.

Formal specifications can help check consistency, but understandably, are not guaranteed to be complete. It is not possible to capture every detail related to a system to make the specification complete. Therefore, often, the assumption is that if a statement is not known to be true in a system, it is assumed to be false. This is called the *Closed World Assumption*. Intuitively, syntactical manipulation, as used in automated reasoning, is not likely to use domain-specific rules and knowledge; hence it is considered a *weak method*. However, there are software programs such as *Expert Systems*, which are based on logic that capture the domain knowledge in a *rulebase* to provide stronger methods of reasoning.

System requirements and descriptions are sometimes ambiguous and inconsistent. Translating them into formal specifications reveal any such imprecision and removes it. Hence the promise that knowledge representation holds, to prevent misinformation from entering the system. Any new statement that needs to be part of the system needs to go through the process of *entailment*, which makes sure that the accepted statement is true. Many companies such as IBM, Microsoft, and Intel have successfully used formal methods for software and hardware verification.

We briefly examined the formal methods-based approaches in Chapter 3. The following sections present a more formal treatise on this topic.

Terminology

We have already used a number of terms from the Formal Methods parlance in the introduction and in the beginning of the chapter. Let us formalize some more terms using more precise and exacting parlance. The terms and their definitions closely follow the real-world phenomena, so they should not need a lot of explanation, as we saw in Chapter 3 and the initial few paragraphs of this chapter. First is the most useful concept of a *proof*. A proof is a form of an argument comprised of a sequence of steps in accordance with a small set of *rules of inference*. The rules help in drawing conclusions from preceding statements. The goal of the proof is to prove that something is true. If the goal fails and it can't be proved to be true, in the Closed-World Assumption (CWA), it is assumed to be false.

The statement that is proved as true by a formal proof is called a *theorem*. Sometimes, the proofs for theorems get really long. In such cases, a *lemma* is used as an intermediate theorem that is used within the proof of another theorem. A theorem often leads to other conclusions called *corollaries*. These are also theorems, which can be established from other theorems that are already proved. In the real world, we find many facts that are self-evident and don't need a proof. These are the fundamental truths to start with. The term used to connote such fundamental truths that are always assumed to be true is *axiom*.

A system is said to be *consistent* if we cannot both prove a theorem and also its converse. An identical concept is *soundness*, which requires that all provable statements be true. Often, research papers are rejected because the work is not sound. In a rush to publish more papers, some researchers try to prove their hypothesis, which may not really be true. Editorial boards return such papers with a comment that the work is *unsound*. An expression that is true in every possible interpretation is called a *tautology*. On the other hand, a statement that is always false is called a *contradiction*. Statements, which do not belong to either categories, are called *contingents*.

As we shall see, theorem proving is so mechanical that constructing the proofs is coded in a computer programming language. Such *Automated Theorem Proving (ATP)* is widely used for a variety of applications in Math, Software design and verification, and Hardware verification. There are many successful ATP systems in use today. There are many other terms, some of which will be introduced as we progress through the chapter. For now, we have enough parlance to start talking about how the veracity problem can be solved when the data we have is represented in logic notation.

Note   Formal methods comprise of a number of well-defined terms, which themselves are stated formally, without any ambiguity.

Propositional Logic

A proposition is a statement that can be true or false. The propositions can be joined by logical connectives. Propositional Logic provides ways of joining or changing propositions to construct other propositions. The common ways of combining natural language sentences by using *and, or* applies to propositional logic as well. Propositions can be combined using *operators* such as *and* and *or*. Also, just like *not* is used to change the meaning of an English sentence, it is used as an operator with the same semantics in Propositional Logic as well. All this sounds like English and other natural languages. What is different about Logic that makes it so useful in dealing with the veracity problem? It is the mathematical abstraction that gives these concepts power to transform the veracity problem into that of mathematical reasoning.

Math requires notation. Table 6-1 shows the syntax (notation) for the operators we discussed in the previous paragraph. A negation operator simply changes the value of a true proposition to false and vice versa.

The semantics for other operators are given in Tables 6-2 to 6-5. The semantic interpretations are quite intuitive and follow the real-world interpretations quite closely. For instance, two true statements joined by an *and* operator produce a true statement. If any of the two statements are false, the resulting statement from an *and* conjunction will be false. For instance, if "John secured A grade in English" and "John secured A grade in Science" are true statements, the combined statement, "John secured C grade in English and A grade in Science" is intuitively false and is false even according to the *truth table* in Table 6-2. The *T* in the table stands for True and *F* for False.

The operators can be used any number of times in a sentence. For instance, $(p \rightarrow q) \wedge (q \rightarrow p)$ is a valid sentence. In fact, if a truth table is drawn for this expression, it can be verified as equivalent to $(p \leftrightarrow q)$. Truth tables are similarly used to determine the equivalence of different expressions. Using truth tables, it can be determined that $(p \vee \neg p)$ is a tautology, while $(p \wedge \neg p)$ is a contradiction that is always false. It can also be determined that $(p \rightarrow q)$ is equivalent to $(\neg p \vee q)$ and $\neg(p \vee q)$ is equivalent to $(\neg p \wedge \neg q)$.

Table 6-1. *Propositional Logic Notation*

Symbol	Meaning
$\wedge$	And (conjunction)
$\vee$	Or (disjunction)
$\rightarrow$	If (implication)
$\neg$	Not (negation)
$\leftrightarrow$	If and only if (equivalence)

Table 6-2. *Semantics for the conjunctive and operator*

Proposition p	Proposition q	p ∧ q
T	T	T
T	F	F
F	T	F
F	F	F

Table 6-3. *Semantics for the disjunctive or operator*

Proposition p	Proposition q	p ∨ q
T	T	T
T	F	T
F	T	T
F	F	F

Table 6-4. *Semantics for the implication operator*

Proposition p	Proposition q	p → q
T	T	T
T	F	F
F	T	T
F	F	T

Table 6-5. *Semantics for the equivalence operator*

Proposition p	Proposition q	p ↔ q
T	T	T
T	F	F
F	T	F
F	F	T

Table 6-6. *Converse, Inverse, and Contrapositive of an implication*

Notation	Meaning
$p \to q$	If p then q
$q \to p$	Converse of the 1st row (if q then p)
$\neg p \to \neg q$	Inverse of the 1st row (if not p then not q)
$\neg q \to \neg p$	Contrapositive of the 1st row (if not q then not p)

Table 6-6 illustrates the semantics of the terms *converse, inverse, and contrapositive*, which are sometimes useful when constructing proofs of lemmas, theorems, and drawing corollaries. The interpretation again is quite intuitive. Suppose *p* is "John gets an A grade" and *q* is "John works hard." Then the contrapositive of their implication is "If John does not work hard, then John does not get an A grade."

As can be seen, the more math we develop around the representation of information, the more tooling we have to determine the truthfulness of complex statements. Truth table is one such construct. However, if there are *n* propositions in an expression, we need 2^n rows in the table and that many computations. Clearly, such a computation becomes intractable. Therefore, most logic systems depend on *inference rules*. Truth can be checked by truth tables, row by row or by using rules of inference. We saw an example of an inference rule in Chapter 3 called *modus ponens*. There are more such rules as tabulated in Table 6-7. Please note that the *c* in the table stands for *contradiction*. If the propositions given in the second column are true, the propositions in the third column can be inferred to be true as well. The given expression can therefore be transformed into the inferred expression.

Note   Inference rules!

Table 6-7. *Inference rules of Propositional Logic; c stands for contradiction*

Inference Rule	Given	Infer
Modus ponens	$p, p \rightarrow q$	q
Modus tollens	$\neg q, p \rightarrow q$	$\neg p$
Generalization	*Either p or q*	$p \vee q$
Specialization	$p \wedge q$	*Either p or q*
Conjunction introduction	p, q	$p \wedge q$
Replacement	$p \rightarrow q$	$\neg p \vee q$
Disjunctive Syllogism	$p \vee q, \neg p$	q
Disjunctive Syllogism	$p \vee q, \neg q$	p
Transitivity	$p \rightarrow q, q \rightarrow r$	$p \rightarrow r$
Contradiction	$\neg p \rightarrow$	p
Division into cases	$p \vee q, p \rightarrow r, q \rightarrow r$	r

Like in the case of modus ponens, the rules are quite intuitive. Take, for instance, the Disjunction Syllogism described in Table 6-7. Say p is "John is intelligent" and q is "John failed the exam." If we know that $\neg p$ and $p \vee q$ are true, which means "John is not intelligent" and "either John is intelligent or John failed the exam," then intuitively, "John failed the exam" must be true. The row for Disjunction Syllogism in Table 6-7 essentially confirms this. Similarly, the rule for transitivity can also be understood intuitively. If John is intelligent, then John will pass the exam and if John will pass the exam, then John will graduate. From the preceding implications, it can be inferred that if John is intelligent, then John will graduate. Table 6-8 gives a list of logical identities where the LHS is identical to RHS. These identities can be verified by constructing truth tables for LHS and the RHS. These logical identities also come in handy when developing proofs.

The rules in Table 6-7 have different names as well. For instance, the *Generalization* rule is also called as *or introduction*, sometimes denoted as $\vee I$ – the first symbol standing for *or* and the second for *Introduction*. That is because we start with a premise and introduce *or* on the RHS. Similarly, *Specialization* is same as *and Introduction*, denoted as $\wedge I$. It must be noted that all the inference rules can be proven to be *sound*.

Note   The semantics required for reasoning are encapsulated in the syntax of propositional logic, so that the task of theorem proving or deducing truth reduces to syntactic manipulations.

Table 6-8. *Propositional Logic Identities*

Identity name	L.H.S.	R.H.S.
Identity	$p \wedge T$	p
	$p \vee F$	p
Idempotent	$p \vee p$	p
	$p \wedge p$	p
DeMorgan's	$\neg(p \wedge q)$	$\neg p \vee \neg q$
	$\neg(p \vee q)$	$\neg p \wedge \neg q$
Commutative	$p \vee q$	$q \vee p$
	$p \wedge q$	$q \wedge p$
Absorption	$p \vee (p \wedge q)$	p
	$p \wedge (p \vee q)$	p
Negation	$\neg(p \rightarrow q)$	$p \wedge \neg q$
	$\neg(p \leftrightarrow q)$	$\neg p \leftrightarrow q$
Distributive	$p \wedge (q \vee r)$	$(p \wedge q) \vee (p \wedge r)$
	$p \vee (q \wedge r)$	$(p \vee q) \wedge (p \vee r)$
Associative	$(p \vee q) \vee r$	$p \vee (q \vee r)$
	$(p \wedge q) \wedge r$	$p \wedge (q \wedge r)$
Double Negation	$\neg(\neg p)$	p
Implication	$p \rightarrow q$	$\neg p \vee q$
	$p \rightarrow q$	$\neg q \rightarrow \neg p$
Biconditional	$p \leftrightarrow q$	$\neg q \leftrightarrow \neg p$
	$p \leftrightarrow q$	$(p \rightarrow q) \wedge (q \rightarrow p)$

Using inference rules, given axioms, and logical identities, the proof proceeds mechanically by substituting the given expression with the inferred expression. Proof thus becomes a series of syntactical substitutions that preserves the truth with every substitution. Let us consider an example problem. In John's school, to enroll in COEN498, he has to complete COEN490 and either COEN200 or COEN220. COEN498, COEN490, COEN200, and COEN220 are different courses offered by the school. We know that John did not complete COEN200 but completed COEN498. John is claiming that he completed COEN220. Is he stating the truth?

The first step in finding the truth using proposition logic is to translate the problem into propositions, using the notation we just learned. We first start with the following propositions:

p stands for "John completed COEN498"

q stands for "John completed COEN490"

r stands for "John completed COEN200"

s stands for "John completed COEN220"

Now, the problem is to prove s given the following premises

$$p \rightarrow q \wedge (r \vee s) \tag{6.1}$$

$$p \tag{6.2}$$

$$\neg r \tag{6.3}$$

We proceed with the proof by starting with the first two axioms and applying substitutions based on inference rules, using the third axiom when it is time. The proof is shown in Table 6-9 and confirms that John's claim is true and he did complete COEN220. As can be seen from the table, the proof is purely syntactic, giving credence to the question we started this chapter with. Starting with the axioms and applying rules successively to arrive at the goal is called *forward chaining*. On the other hand, starting with the goal and working backward to the axioms is called *backward chaining*. Both algorithms can be proven to be *sound* and *complete* and are of the order $O(n)$, where n is the size of the *knowledge base* comprising the set of given premises.

Note   Forward chaining is data driven, while backward chaining is goal driven.

Table 6-9. Proof that John completed COEN220

Sequence #	Transformation	Rule Applied
1	$p \rightarrow q \wedge (r \vee s)$	Given axiom
2	p	Given axiom
3	$q \wedge (r \vee s)$	Modus Ponens on 1 and 2
4	$r \vee s$	Specialization
5	$\neg r$	Given axiom
6	s	Disjunctive Syllogism on 4 and 5

In the above example, we started with a number of premises, (6.1), (6.2), (6.3) and from them, we *inferred s*. In general, when the data is represented using logic notation, the veracity problem can be transformed into that of concluding s given a number of premises, $p_1, p_2, p_3, ..., p_n$. This is written as:

$$p_1, p_2, p_3, ..., p_n \vdash s \tag{6.4}$$

Expressions like (6.4), where a conclusion is arrived at from a set of given premises is called a *sequent*. The logic system used for such proofs is called *sequent calculus*.

A highly successful language that uses some of the concepts discussed above is *Prolog*. A Prolog program contains a user-provided Knowledge Base (KB) comprising a set of facts and rules. Prolog's inference engine is primarily based on just one inference rule – modus ponens. Using the KB provided by the user and the inference engine, Prolog is able to draw conclusions and determine if a claim is true or false. This is illustrated in Figure 6-1. The KB facts and rules are in the form of *clauses*. Clauses are *literals* connected using disjunctions (*or* operators). Literals are atomic propositions or their negations. Facts and rules in a Prolog program are called *Horn Clauses*, named after the logician, Alfred Horn.

Figure 6-1. *Determining truth in Prolog*

Prolog is used today even in unexpected ways in applications such as the code review tool, Gerrit. It was widely used in Europe and Japan, particularly as part of the Fifth-Generation project in the 1980s and early 1990s. A Prolog program is a set of Horn clauses with some value-adds such as support for arithmetic functions and constructs such as a *cut* to prevent *backtracking*. The evaluation of the truthfulness of a new claim proceeds by backward chaining that was described earlier. In the process of backward chaining, backtracking is the process of going back to the previous split to see if there are other solutions, when the current subgoal fails.

The goal search in Prolog proceeds in a depth-first fashion, left to right, expressed by the recursion in the following two clauses.

$$dfs(X):-goal(X)$$

$$dfs(X):-successor(X,Subgoal),dfs(SubGoal)$$

The Prolog operator *:-* stands for *if*. In the above lines, *X* is a variable. Given a goal to be proved true or false, Prolog proceeds in the above recursive fashion. The first line is the termination condition of the recursion. If we already have the premise that the variable *X* stands for the knowledge base, the first condition matches and the program terminates with a *True* return value for the given goal. Otherwise, the second rule kicks in. Prolog proceeds to try the goal by traversing the knowledge base in a depth-first search fashion. In accordance with the Closed-World Assumption (CWA) discussed earlier, if Prolog cannot prove the goal, it is assumed to be false.

Note   Logic amazingly converts the problem of verifying truth to mechanical manipulation of symbols.

Propositional Logic is also called *Zero–order Logic*. Its expressiveness and capabilities are limited. *Higher Order Logic (HOL)* systems are more expressive and more powerful. For instance, the *First Order Logic* (FOL) allows *quantification* over variables, *Second Order Logic* allows quantification over sets of variables, *Third Order Logic* allows quantification over sets of sets. and so on. Using quantification, we can do operations with quantities, such as all instances or at least one instance of an entity. For most purposes, First Order Logic, also called *First Order Predicate Calculus* or simply Predicate Calculus is usually sufficient within the complexity limits. We will examine this important order of logic that is the basis for powerful logic systems in the next section.

Predicate Calculus

We have not so far used the concept of a variable. Every entity was a constant. Imagine math without the notion of a variable! We still did quite a lot without using any placeholder variables. When we talk of Big Data, we should be able to represent many variants and therefore the need of a variable. Real-world data deals substantially with quantities of entities, which also was not used until now. Combining them both, First Order Logic (*FOL*) provides for *quanitfication* on variables that was not possible with Propositional Logic. The quantity aspect of variables is often expressed in a natural language such as English using phrases such as *there exists* and *for all*. These phrases are indicative of the quantities of entities; hence they are called *qunatifiers* in logic.

Propositions were either true or false. Their truthfulness did not vary. For instance, the proposition, "*Mary* is the mother of *Ann*" was either true or false. If it was true, it remained true at all times. A *predicate*, such as "*x* is the mother of *Ann*, on the other hand can be true or false and change that value depending on the value of the variable *x*. Here, *Mary* and *Ann* are *constant symbols*, real entities of the world. A predicate symbolizes a *relation*. The previous predicate can be symbolically written as

$$Mother(x, Ann) \tag{6.4}$$

In the above predicate in (6.4), *x* and *Ann* are called *terms* and denote objects. The above predicate is *binary* because it takes two terms. Predicates can be combined using the operators we used in Propositional logic. For instance, the *sentence* in (6.5) could mean Mary is Ann's mother and Ann is Coco's mother.

$$Mother(Mary, Ann) \land Mother(Ann, Coco) \tag{6.5}$$

$$Mother(Mary, Ann) \rightarrow Female(Mary) \tag{6.6}$$

The sentence, (6.6) uses the if operator to combine the predicates and means, "If Mary is the mother of Ann, then Mary is female." As can be seen, forming the sentences in First Order Logic plays a crucial role in representing data and reasoning with it to determine the veracity of a claim.

As mentioned earlier, FOL provides two more operators, the *Universal Quantifier* ∀, meaning "for all" and the *Existential Quantifier* ∃, which means "there exists." Using the universal quantifier, (6.6) can be generalized as in (6.7) and (6.8) below. (6.7) can be interpreted as, "For all entity x, if x is a mother of any other entity y, then x is a female." On the other hand, (6.8) can be interpreted as, "There exists an entity x, such that if another entity y is living, then x is the mother of y."

$$\forall x Mother(x, y) \rightarrow Female(x) \tag{6.7}$$

$$\exists x IsLiving(y) \rightarrow Mother(x,y) \tag{6.8}$$

From a veracity perspective, we are mostly concerned with inferencing truth. New connectives will require new inference rules. Just like in propositional logic, we have *introduction* and *elimination rules* for the two quantifiers. Also like in propositional logic, these rules can be proven to be *sound*. Let us start with the easiest one first – the elimination of the Universal quantifier. We can eliminate the ∀ operator by *instantiating* the variable with a specific instance. For instance, in (6.7), we can instantiate x with a specific value, *Mary*, and eliminate the Universal Quantifier. The resulting sentence is given in (6.9) below.

$$Mother(Mary, y) \rightarrow Female(Mary) \tag{6.9}$$

In (6.9), we substituted x with a constant, *Mary*. We can also instantiate the variable with a *function*, such as *MotherOf(Ann)*, which looks quite like a predicate but is very different. A predicate is either true or false and nothing else. On the other hand, a function returns a value in a domain. For instance, *MotherOf(Ann)* may return *Mary*, whereas *Mother(Mary,Ann)* can return only true or false. We can also eliminate the Universal quantifier from (6.7) by transforming it into (6.10) below.

$$Mother(MotherOf(Ann), y) \rightarrow Female(MotherOf(Ann)) \tag{6.10}$$

It must be noted that the instantiation must happen uniformly throughout the sentence. So, in (6.9) we cannot instantiate x with *Mary* on the Left-Hand Side (LHS) and some other value on the Right-Hand Side (RHS). The Universal quantifier elimination rule is denoted as $\forall E$.

The next easier rule is *Existential Introduction,* $\exists I$, sometimes called *Existential Generalization*. It is easy because if a sentence is true for an instance, then we can generalize that there *exists* an entity such that the sentence is true. Using this rule, (6.9) can be generalized as (6.11) below.

$$\exists x Mother(x, y) \rightarrow Female(x) \qquad (6.11)$$

We substituted *Mary* with the variable x in the sentence, (6.9) and introduced the existential quantifier.

Constructing proofs in First Order Logic sometimes requires introduction of the Universal quantifier, denoted as $\forall I$. The goal here is to transform a sentence like (6.10) to a generalized form like (6.7). This can be done only if it can be proven that there is nothing specific to the value, *Mary* in (6.9) that makes it true only for that instance and that the sentence is true even for other values that can be substituted in place of *Mary*. In case of (6.10), this can be easily proved because (6.10) is a result of Universal elimination from (6.7). Universal introduction into a sentence can therefore easily happen after a series of sentences following a Universal elimination from the same sentence.

Now for the most difficult of the rules, the *Existential Elimination,* $\exists E$. It can be used to convert sentences such as (6.11), which includes an existential quantifier to (6.9) that eliminates the quantifier, by substituting a specific value of the variable x. But how do we know that the specific value instantiated makes the sentence true? Existential elimination can therefore only happen when it can be proven that the specific instance satisfies the sentence, when substituted for the variable, throughout the sentence.

Let us consider an example. We want to know whether the teacher's claim that "A student who has graduated secured an A grade" is true or not, given the facts that "A student in the class secured an A grade" and "Every student in the class has graduated." As before with the propositional logic example, we first frame the following premises:

Student(x) stands for "x is a student of the class"

Agrade(x) stands for "x secured A grade"

Graduated(x) stands for "x graduated"

The domain, more formally called the *domain of discourse or universe of discourse* here is the set of all students in the class. The given premises can be written as (6.12) and (6.13) below.

$$\exists x \big(Student(x) \wedge Agrade(x) \big) \tag{6.12}$$

$$\forall x (Student(x) \rightarrow Graduated(x)) \tag{6.13}$$

The teacher's claim that needs to be evaluated for its veracity can be written as (6.14) below.

$$\exists x \big(Graduated(x) \wedge Agrade(x) \big) \tag{6.14}$$

From the proof in Table 6-10, it does appear that the teacher's claim in (6.14) is indeed true.

***Table 6-10.** Proof that the teacher's claim is true*

Sequence #	Transformation	Rule Applied
1	$\exists x \big(Student(x) \wedge Agrade(x) \big)$	*Given premise* (6.12)
2	$\big(Student(joe) \wedge Agrade(joe) \big)$	$\exists E$ *from #1, assuming joe satisfies* (6.12)
3	*Student (joe)*	*Specialization of # 2 above*
4	$\forall x(Student(x) \rightarrow Graduated(x)$	*Given premise* (6.13)
5	$Student(joe) \rightarrow Graduated(joe)$	$\forall E$ *from #4 above*
6	*Graduated (joe)*	*Modus ponens on # 3 and # 5*
7	*Agrade (joe)*	*Specialization from # 2*
8	$Graduated(joe) \wedge Agrade(joe)$	*Conjunction introduction into# 6 and # 7*
9	$\exists x \big(Graduated(x) \wedge Agrade(x) \big)$	$\exists I$ *into #8 above*

Note   Logic is the language of truth. Truth or the lack of it reveals itself, when subjected to logic.

Fuzzy Logic

The foregoing discussion assumed that data can be represented precisely without any ambiguity or uncertainty. However, such precision is too exacting a demand on the real world – we cannot always talk in absolute terms of true and false or 1 and 0. In the Twitter example, we saw in the chapter on Machine Learning, Chapter 5, we cannot express the dependencies with certainty. For instance, we cannot conclude that a tweet is true if it has multimedia content. We can only conclude that there is a high *chance* that it is true. Obviously, such an imprecise domain of discourse cannot be modeled in Boolean Logic, where only two outcomes, true or false, are possible. Also, Machine Learning algorithms such as Neural Networks can become compute intensive and intractable. We clearly need a way to reason under circumstances of uncertainty.

Fuzzy Logic, invented by Lofti Zadeh in the 1960s, extends the Boolean Logic concepts to cover imprecision as well. It is much closer to the real world and how people think. Boolean Logic is in fact a boundary case of Fuzzy Logic. Instead of precise modeling, we have elastic constraints defining the scenario. These elastic constraints propagate through the inference process. The elastic constraints are subjective, based on opinions rather than on frequencies or precise measurements. The boundaries between classes are soft and the transition between them is smooth. A good analogy is that of Professors sometimes going soft on students by giving some credit to the answers instead of just giving a zero score. Techniques such as Fuzzy Logic, where the class boundaries are soft, come under what is known as *soft computing*.

Note   Fuzzy Logic models the real world much more closely than the idealistic Boolean Logic.

In classical set theory, the membership of an element is *crisp* – either it is present in the set or not. In Fuzzy theory, the element's membership in a set A is *fuzzy*, given by a *membership function,* μ_A that takes values between 0 and 1. The membership function indicates the *degree* to which an element belongs to a set. Figures 6-2 and 6-3 show two kinds of membership functions for the tweets example we discussed in Chapter 5. The membership function plots the membership of each tweet in three

classes (regions bounded by thin, thick and dotted lines in the figure):
a) False, b) True, c) Uncertain, based on the number of retweets that it
received. As can be seen, the boundaries between these classes is not crisp.
In fact, there is appreciable overlap between the classes. That means that
tweets having a certain number of retweets can be true, false, or we do not
know if they are true or false. This scenario is much closer to the real-world
situation. Just because a tweet has a certain number of retweets does not
mean it is either true or false. Thus, we have successfully modeled a real-
world situation that would not have been possible with Boolean Logic.

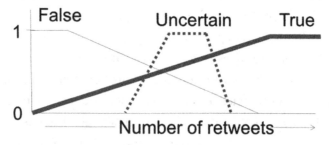

Figure 6-2. *Trapezoidal Membership Function*

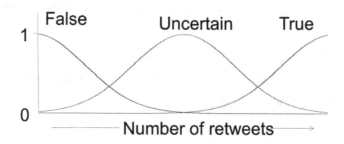

Figure 6-3. *Gaussian Membership Function*

   The membership function in Figure 6-2 is trapezoidal, with parallel
horizontal sides. But from probability and statistics, we know that data
in a number of domains of discourse is *normally* distributed and is
represented by *Gaussian* functions of the form $e^{-\beta x^2}$, where β is a positive

real number. For instance, the probability density function of a normal distribution is given by the Gaussian function, $\dfrac{1}{\sqrt{2\pi}\sigma}e^{\frac{-(x-\mu)^2}{2\sigma^2}}$, which is of the form of the latter. Therefore, often, the fuzzy membership function takes the form of a Gaussian function and the plot looks similar to the one in Figure 6-3.

As another example, the membership function of a fuzzy set of real numbers *almost* equal to 1 can be modeled by the Gaussian function given in (6.15) below.

$$\mu_X = e^{-\beta(x-1)^2} \tag{6.15}$$

It must be noted that the membership function can take many shapes. Figure 6-2 and Figure 6-3 are two of those shapes, probably the more popular ones. The trapezoidal membership function has the values depicted in Table 6-11 given below.

Table 6-11. *Value of a trapezoidal membership function*

Value of μX	When value of x is
$1 - \dfrac{(a-x)}{\alpha}$	$a - \alpha \leq x \leq a$
1	$a \leq x \leq b$
$1 - \dfrac{(x-b)}{\beta}$	$b \leq x \leq b + \beta$
0	*all other values*

Using the notion of a membership function, Propositional Logic and Predicate Calculus can be extended to model fuzziness. The definitions of the operators change as shown in Table 6-12, to incorporate the fuzziness. In the example given in the table, assume T is the set of true tweets and F is the set of false tweets. The membership function μ_T gives the possibility that a tweet is true and μ_F indicates the possibility that a tweet is false. As can be seen, they add up to more than 1.0, indicating the *granulation* and *graduation* of the fuzzy sets. Granulation refers to the fact that there is a range associated with a class and not just a single value. Graduation implies that the transitions between the classes are gradual, with some overlap.

Table 6-12. *Fuzzy interpretation of the logical operators*

Operator	Meaning	Fuzzy Interpretation	Example ($\mu_T = 0.7$, $\mu_F = 0.4$
$\wedge$	And (conjunction)	$\min(\mu_T, \mu_F)$	0.4
$\vee$	Or (disjunction)	$\max(\mu_T, \mu_F)$	0.7
$\neg$	Not (negation)	$1 - \mu_T$	0.3

The interpretations in Table 6-12 can be verified to be true by limiting μ_T and μ_F to only 1 or 0 and cross-checking with the Boolean Logic interpretation. For instance, when $\mu_T = 1$ and $\mu_F = 0$, the conjunction results in 0 because $\min(\mu_T, \mu_F) = 0$, while the disjunction results in a 1. The laws of commutativity, associativity, distributivity, idempotency, and De Morgan's Laws, given in Table 6-8 in the section on Propositional logic all apply in Fuzzy Logic as well. The inference mechanisms of Boolean Logic can similarly be extended to handle imprecision in Fuzzy Logic.

A useful metric for Fuzzy sets is its *entropy*. It is defined in (6.16) below.

$$Entrooy(T) = \frac{|T \cap T^c|}{|T \cup T^c|} \tag{6.16}$$

Entropy is an indication of the disorder or imprecision in the class. For a crisp set and Boolean Logic, it can be easily seen that Entropy is zero, because $\left|T \cap T^c\right| = 0$.

Note Though invented much earlier in time, the classical Boolean Logic can be seen as a special case of Fuzzy Logic.

Summary

Representing data in appropriate ways can facilitate reasoning with it to determine its truthfulness. In this chapter, we examined how the classical Boolean Logic and Fuzzy Logic can be used to model the real-world and draw inferences. We used the same Twitter example we used in Chapter 5 to demonstrate how Fuzzy Logic can help deal with entropy in the truthfulness of the tweets. The techniques presented in this chapter call for a change in the representation of Big Data. There is plenty of scope for further evolving this topic to cover various forms of Big Data. In the next chapter, we shall examine some more methods, particularly from the Information Retrieval area, to address the veracity problem.

EXERCISES

1. Draw a table depicting the semantics of a contrapositive of an implication. Compare the table with that for the implication.

2. Recollect some real-world scenarios where people were tricked into believing something because they could not reason and see through the logic clearly. Construct proofs such as the one in Table 6-10 to prove the contrary claim in such scenarios.

3. In this chapter, we examined a few logic operators such as →, ¬, ∧, ∨, ∃, ∀ . We studied how the last two operators, the quantifiers, were a significant value-add to the what was possible in Propositional Logic. Can you think of any other operators, the introduction of which can substantially enhance the ability to model the real world in Boolean Logic? If so, provide their syntax, semantics, and a discussion on how they help further.

4. What are some of the real-world sets that can be modeled as Fuzzy sets? Describe the characteristics of their membership functions, preferably in equation form such as in (6.15) or Table 6-11.

CHAPTER 7

Medley of More Methods

Big Data comes in many types – text, video, audio, structured, unstructured, legacy, data from newer sources, and so on. The variety aspect that is manifest in Big Data characterizes the veracity problem as well. We therefore need a variety of methods to solve the various manifestations of the veracity problem. The broad categories of these approaches are covered in the remaining chapters. This chapter is meant for the methods that do not fall in any of the other broad categories described in this book and cannot be generalized into a single category. Hence, we call the collection of these methods a medley. We briefly introduced them in Chapter 3. In this chapter, we shall look at them in more detail.

Collaborative Filtering

Rumor mills thrive on informal collaboration. Rumors pass from one to another informally and intensify in the process. In a similar fashion, in formal settings, truth evolves through collaboration as well. Google scholar's slogan reads, "Stand on the shoulders of giants," indicating the highly collaborative nature of research work. Research scholars collaborate with each other using platforms such as Google scholar, to discover the subtlest of the truths. In most formal settings, truth and work based on

© Vishnu Pendyala 2018

V. Pendyala, *Veracity of Big Data*, https://doi.org/10.1007/978-1-4842-3633-8_7

truth evolves through collaboration. People come to conclusions about the truthfulness of claims in correlated ways. We can often guess a person's actions based on the action of similar people. Collaborative Filtering (CF) is a mathematical, programmatic abstraction of these concepts. It leverages people's judgment to deal with uncertainty and incomplete information. Collaborative Filtering is a major area, using a number of probabilistic and non-probabilistic algorithms. This chapter introduces the topic using non-probabilistic methods.

Collaborative Filtering is primarily used to recommend items based on the personal tastes of similar users and similar items. This chapter extends those ideas to the veracity domain. Using CF in the veracity domain can possibly help search engines to provide results that are more attuned to the beliefs, values, and intuition of the user. Search engines can assign a truthfulness index from the user's perspective. A Question Answering (QA) system can determine the truthfulness of a claim made by a user from the perspectives of similar users. Going back to our microblogging example, it can be determined if a user's account is hacked based on the changes in the correlation coefficient of the user's posts with respect to those of other users. Microblogging users can find other users with similar beliefs as them. For instance, to a user who posted that global warming is true, Twitter can suggest other users who believe in a similar level of truth in global warming, for future collaboration and interaction.

Note   Owing to the interactions, upbringing, culture, and other social reasons, personal judgments are often correlated.

Collaborative Filtering is the process of filtering information based on human judgments. Suppose there are n claims, $v_1, v_2, ..., v_n$. The claims have been judged by m human beings or users, $u_1, u_2, ..., u_m$ as likely true by assigning ratings $r_1, r_2, ..., r_n$. For instance, user Joe, referred to as u_1, thinks that the claim v_1, that "Global warming is real" is 90% true, so he gives a rating r_1 of 9 on a scale of 10 to this claim. Another user Don, u_2, does not believe in global warming at all, so he gives a rating r_2, of 0. A third user Bill, u_3, thinks quite along the lines of u_1 and rates the claim with a 9 as well. It is often the case that some of the ratings may be missing because of insufficient data. Collaborative Filtering tries to predict these missing ratings in mathematically rigorous ways.

When we have all the ratings, the claims can be concluded as true or false by statistically processing the data. Table 7-1 illustrates the problem that CF tries to solve, using the preceding simple example. The three users rate three different claims as shown in the table, with the exception that we still do not have the rating that user u_3 will give for the claim, v_3. The problem of CF is to find out this missing rating, from what can be thought of as a *truth matrix*. Intuitively, from the table, the first and third users, Joe and Bill, seem to judge alike, so Bill is more likely to rate the third claim as a 7 like Joe. Collaborative Filtering concludes the same mathematically, by using similarity measures.

Table 7-1. *The Problem of Collaborative Filtering: Truth Matrix*

	v_1: Global Warming is real	v_2: Flying cars will appear by 2020	v_3: Einstein was a flirt
u_1: Joe	9	7	7
u_2: Don	0	9	9
u_3: Bill	9	7	?

Note   The ratings may be explicitly given by the users or concluded from their online activity, using Big Data analytics.

Collaborative Filtering can be user based or claim based. The example in the previous paragraph and illustrated in Table 7-1 is user-based CF. The idea behind user-based CF is that the users who judge alike will continue to do so at all times and on other claims as well. On that premise, we predict a user's rating of the truthfulness of a claim based on the ratings of who judged similarly. In statistics, similarity is measured by correlation. There are various measures for statistical correlation, but the essential idea is to quantify the overlap of judgments between two users. A simple measure of similarity or the lack of it could be the average absolute difference between the ratings of the two users, expressed in the following equation (7.1). In the equation, n is the number of claims that have judgmental ratings of u_1 and u_2 and $r_i(u_1)$ and $r_i(u_2)$ are the ratings given by the two users respectively for the ith claim, which are rated by both users. This is really a distance measure, inversely proportional to similarity. It indicates the lack of similarity.

$$dist(u_1,u_2) = \frac{\sum_i |r_i(u_1) - r_i(u_2)|}{n} \qquad (7.1)$$

An improvement over the above measure in (7.1) is called *Pearson's Correlation Coefficient*, given in (7.2) below. In the equation, $\bar{r}(u_1)$ and $\bar{r}(u_2)$ are the average of all the ratings given by the respective users to various claims.

$$sim(u_1,u_2) = \frac{\sum_i (r_i(u_1) - \bar{r}(u_1))(r_i(u_2) - \bar{r}(u_2))}{\sqrt{\sum_i (r_i(u_1) - \bar{r}(u_1))^2 \sum_i (r_i(u_2) - \bar{r}(u_2))^2}} \qquad (7.2)$$

The value of the Pearson's Correlation Coefficient ranges between -1 and +1. A value of +1 implies perfect alignment between the users, while a -1 implies a completely opposite alignment. These ideas are illustrated in Figure 7-1. A value of 0 for the Pearson's Correlation Coefficient implies no correlation between the users. Since we have a dataset to start with and train our model of prediction, comparing with the Machine Learning approaches, CF is more like supervised learning.

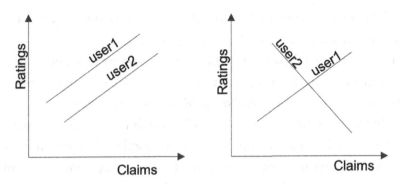

Figure 7-1. *Users Correlation: The plots of the ratings given by similar users (left) are parallel and dissimilar users (right) are orthogonal.*

Note   To those who are familiar with vector algebra, the equation (7.2) can be recognized as similar to the one for the cosine similarity between two vectors.

Using (7.2), we can compute the closeness of any two users. We usually choose k closest users to a given user, to predict the given user's opinion. The value of k is often determined empirically, balancing the tradeoff between the computational intensity resulting from a higher value for k and the suboptimal accuracy from a lower value for k. This approach is called the k-Nearest Neighbors or simply, k-NN algorithm. Using (7.2) and the k-NN algorithm, the rating of a given user, u for a given claim, i can be predicted by the formula given in (7.3) below.

$$r_i(u) = \bar{r}(u) + \frac{\sum_{x \in NN(u)} sim(u, x) \cdot \left(r_i(x) - \bar{r}(x) \right)}{\sum_{x \in NN(u)} sim(u, x)} \tag{7.3}$$

In the above equation (7.3), x denotes one of the k nearest neighbors of user u and $sim(u, x)$ is obtained from (7.2) for each x.

So far, we have only seen user-based CF. Sometimes, a new user may not have given enough ratings to determine his k nearest neighbors. This is known as the user *ColdStart problem*. We can use claim-based CF in such cases. Equations (7.2) and (7.3) remain the same and we can still use the k-NN algorithm with claim-based CF. Just that instead of similarity between users, we compute the similarity between claims. Like in the case of user-based CF, in claim-based CF too, two claims are similar if their ratings are similar. This made semantic sense in case of user-based CF, in that two users who have similar opinions can indeed be similar in thinking. But this may not always make sense in the case of claim-based CF.

For instance, two claims, "Global warming is real" and "Einstein was a flirt" may not be similar in meaning at all. But if many users rate them similarly, for the purposes of claim-based CF, they are treated as similar. We look at the ratings given by a particular user for k claims.similar to the given claim and use similar equations as (7.2) and (7.3) given below to predict the rating for a new claim. For instance, referring to Table 7-1, "v_2: Flying cars will appear by 2020" and "v_3: Einstein was a flirt" have similar ratings. Intuitively, it can be predicted that the missing rating for "v_3: Einstein was a flirt" will also be same as that for v_2, which is 7. The following two equations will also provide the same result.

$$sim(i,j) = \frac{\sum_u \left(r_i(u) - \overline{r}_i\right)\left(r_j(u) - \overline{r}_j\right)}{\sqrt{\sum_u \left(r_i(u) - \overline{r}_i\right)^2}\sqrt{\sum_u \left(r_j(u) - \overline{r}_j\right)^2}} \tag{7.4}$$

$$r_i(u) = \overline{r}_i + \frac{\sum_{x \in NN(i)} sim(i, x) \cdot \left(r_x(u) - \overline{r}_i\right)}{\sum_{x \in NN(u)} sim(i, x)} \tag{7.5}$$

In the above equations (7.4) and (7.5), the average rating of the claim i is given by $\bar{r}_i$, claim j is $\bar{r}_j$ and claim x is $\bar{r}_x$. In (7.4), the summation iterates over the users who rated both claims i and j, while in (7.5), over the k similar claims. If the user is relatively new and has rated only one similar claim, then k is 1.

Note   Similarity in Collaborative Filtering is a highly simplified abstraction of its real-world interpretation.

Vector Space Model

Along the lines of CF, another model that works based on similarities is the Vector Space Model. It can be used to group documents and determine their veracity based on their similarity to known true or fraudulent documents. It can be used to detect plagiarism based on the similarity metrics. The concepts of Vector Space Model are quite simple, but the impact is profound. As in the case of CF, the primary use of the Vector Space Model (VSM) is not in the veracity domain. Apart from plagiarism, VSM has also been used to detect truthfulness at the rhetorical structure level. We therefore examine the concepts involved in VSM in this chapter. Some knowledge of vector algebra may help better understand this topic but is not necessary. Vector can be thought of as a point with a given direction and distance from the origin. Other concepts are explained as they are introduced.

As we saw in Chapter 3, the idea of VSM is represented by each document as a vector in a multidimensional space, with each axis representing a word in the corpus. Let us start with a simple example. Suppose our corpus has just three words: vector, space, and model. Figure 7-2 plots these words in a three-dimensional space with the x-axis representing the word "vector," y-axis representing the word "space," and z-axis representing the word "model." Three documents named Doc 1, Doc 2, and Doc 3 are shown in the figure. The documents contain "Space Model," "Vector Space," and "Vector Model" respectively. Since each word occurs at most once in a document, the corresponding vector's value along the axis of the word is 1 if the word is present in the document and 0 if the word is not present. As shown in the figure, Doc 1 therefore is represented by the vector with coordinates [0,1,1], Doc 2 by the vector with coordinates [1,1,0], and Doc 3 by [1,0,1].

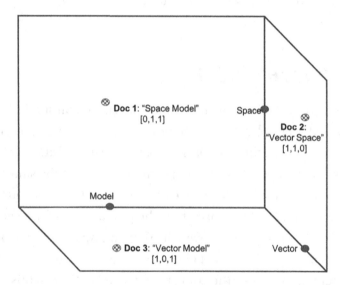

Figure 7-2. *The Vector Space Model of a three-word Corpus*

The above is a simple scheme of a limited corpus represented by the Vector Space Model. In a typical corpus with thousands of words, the dimensions are also in several thousands. Each document is modeled as a vector in this multidimensional space. Instead of just the count of the words, as we saw in Chapter 3, the value along each dimension is the TF. IDF value. As you may have understood by now, we do not care for the syntactic or semantic structure of the document. We treat documents as a bag-of-words. Hence, VSM is a *Bag – of – words* representation. Once the documents are represented as vectors, finding similarity between documents is a matter of applying vector algebra operators. A simple Euclidean distance or simply distance may not be very effective. This can be illustrated by an example. Suppose the document d1 is appended to itself to produce d2. The Euclidean distance between the corresponding vectors will be long even though they are similar.

A better measure for similarity is to measure the angle between the vectors. In the above example, the angle between d1 and d2 is 0. Hence, we often use the trigonometric ratio called Cosine of the angle between the vectors to measure similarity between the corresponding documents. It must be noted that cosine is a monotonically decreasing function in the [0,180] degrees interval. That means, the bigger the angle, the lesser is the cosine value and the more dissimilar the documents are. Cosine similarity is illustrated in Figure 7-3. Let us call the document in question "Query." Its similarity to document called Doc 6 is illustrated in the figure. Once again, for better visualization, the vectors are plotted in a three-dimensional space, but in reality, there could be thousands of dimensions – as many as there are unique words in the corpus. The three-dimensional visualization can be thought of as an orthogonal projection.

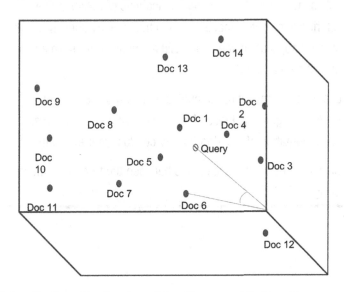

Figure 7-3. *Cosine Similarity of the Query with Doc 6*

Similarity metrics such as cosine value of the angle between the vectors can indicate plagiarism. In our microblogging example, if the document containing the text of a user's posts over a period of time is quite different from the document containing posts on the same topic by other users until then, as evidenced by a low cosine similarity value between the two documents, there is quite a chance that the user is fake or has a personal agenda to drift the influence.

Summary

In this chapter, we examined two methods based on similarity measures, which can help in the veracity domain. On to the next chapter with an exciting discussion on what is being touted as the 21st century's revolution in trustless computing – the blockchain.

EXERCISES

1. This chapter listed only a few applications of Collaborative Filtering and Vector Space Model in the Veracity domain. Think of other application scenarios. List as many as possible and describe them briefly.

2. Examine the impact of the choice of k on the accuracy of the k-NN algorithm. Please feel free to use an open source implementation of the algorithm to conduct your experiments.

3. Calculate the cosine similarities between the three documents in Figure 7-2.

CHAPTER 8

The Future: Blockchain and Beyond

According to a popular saying in Sanskrit, the basis or root of this world lies in money. When the world trusts its money only with a technology – neither backed up by the government nor by people, the technology ought to be entirely trustable. Blockchain is that technology, which is revolutionizing the way the world transacts, enhacing the veracity aspect substantially. Cryptocurrency, which is based on the Blockchain technology, is rapidly gaining acceptance everywhere. Bitcoin, which is one of the popular cryptocurrencies, recorded a market cap of over 300 billion USD toward the end of 2017, comparable to that of companies like Facebook. The market cap of all the cryptocurrencies, including Bitcoin, at the beginning of 2016 was around 7 billion USD. Clearly, the rapid growth is an astounding testimony to the promise the technology holds. Given that cryptocurrencies lack a government's support or backing, the trust they enjoy is almost entirely attributable to the underlying technology – Blockchain.

In Chapter 6, "Formal Methods," we explained how changing the underlying representation of information into logic can improve its verifiability and therefore veracity. On one hand, it was a fundamental shift in the way applications use Big Data. Blockchain, on the other hand can enable a fundamental shift in the way Big Data is generated. When the sources generating the Big Data are tracked using the Blockchain technology, the authenticity of the data is substantially enhanced.

© Vishnu Pendyala 2018
V. Pendyala, *Veracity of Big Data*, https://doi.org/10.1007/978-1-4842-3633-8_8

Blockchain provisions a *trustless network*, where participants do not trust each other, but still transactions among them can be fully trusted because of the underlying technology.

Suppose there is a small world – a closed world, where any new information is considered true only if everyone in the world agrees that it is true. Everyone knows every single truth in this small world. Then there is no place for lies. Blockchain tries to build this ideal world, entirely digitized. This concept is similar to the CWA – Closed World Assumption that we saw in Chapter 6, "Formal Methods." We saw how Prolog programs contain a Knowledge Base (KB) of facts and rules and any new sentence will have to align with the existing KB, something that is verified by the process of inferencing. A new claim cannot enter the KB if it does not agree or entail from the facts currently existing in the KB. Figure 8-1 illustrates this concept. Every participant in the closed world knows all the truths, and a claim is true only when there is consensus.

Note   Blockchain is an entirely technological solution to solve a predominantly social problem of mendacity.

Blockchain is not a logic-based approach though. Instead, Blockchain uses Database, Cryptography, and P2P networks to build a closed world of truthful, trusted, and tamper-proof system for transactions and the massive amounts of data they generate. Blockchain helps in reducing the need for middlemen such as notaries to enlist trust in transactions or a central authority to certify their validity. It has many other advantages such as improving the transaction efficiencies, imparting transparency, and a potential to automate payments. The significant contribution of the technology is still to the veracity of data that is generated through the cycle of any transaction. Transaction data is a significant part of the Big Data metamorphosis. Given that most of the Big Data Analytics run on transaction data, Blockchain can usher in an era of uncompromised quality of data.

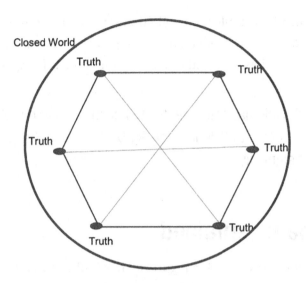

Figure 8-1. *Consensus-based truth in a closed world*

Blockchain's popularity owes to its use in cryptocurrencies. Blockchain filled many gaps in the traditional financial systems. It eliminated the need for centralized control; and reduced inefficiencies, costs, and vulnerabilities from the system. Quite a few Blockchain concepts are borrowed from the finance world. Veracity of financial accounting revolves around the concept of the *ledger*. Money flows and *asset* movements are meticulously tracked in the ledger. Every entry in the ledger is a *transaction*. The rules governing the transactions constitute a *contract*. Any subsequent information can be tallied against the ledger to confirm or dispute.

The ledger forms the basis of truth, just like the Knowledge Base in Formal Methods. Thus far, the ledger has been mostly used for the recordkeeping of financial information. It has been maintained centrally by trusted authorities like banks and other financial institutions. Blockchain makes the ledger much more ubiquitous and expands its scope enormously, particularly to the transactions of the *Internet of Things* (IoT). Every participant of the "closed world" we discussed earlier has a copy of the same ledger. Blockchain in that sense can be described as a *distributed ledger*. The ledger can be thought of as a database without a modify function. It has a trusted trail of anything that is being tracked.

Note   Blockchain has the potential to become the immutable, tamper-proof, ubiquitous ledger for Big Data, solving the veracity problem substantially.

Blockchain Explained

The simple scheme depicted in Figure 8-2 has problems. How do we make sure that one party in the closed world maliciously disagrees with a transaction and does not allow consensus to ensue? What will prevent someone from changing the truth forgotten by all others? What if one of the participant refuses to accept new data? We therefore need to tighten up the design by adding more requirements. Some of these requirements could be the following:

- Transactions should work in a trustless environment, where no one trusts anyone else, but still be entirely trustworthy.

- Data with each participant must be immutable and identical to that with other participants, which calls for *collective bookkeeping* and transparency.

- There must be consensus among the participants, if the data in question is true.

- For widespread application, data must be general enough, not specific to any unit of measurement.

- To expand the usage, the participants need not necessarily be human beings.

A few implementation imperatives emerge from the preceding requirements:

- Information with each participant must be digitized to make it a digital ledger.

- The process should be entirely automated, without the human element, as much as possible, so that there is no subjectivity.

- To make sure that the data is tamper-proof, we need to involve cryptographic checksums, or better still, *cryptographic hash functions*, explained in the box below.

- Data must be recorded chronologically and to preserve the chronological order, data chunks must be linked as in a chain.

- There must be replication processes to make sure that the digital ledger copies distributed across the participants are identical.

- To prevent frivolous attempts at hacking and adding fraudulent blocks of information, the process of adding blocks must be non-trivial.

- Since we are dealing with Big Data, we need efficient data structures to process information in ways that reduce the bandwidth, storage, and compute resources.

Note   The novelty of the Blockchain solution is in bringing together a host of known solutions to solve an important problem.

CRYPTOGRAPHIC HASH FUNCTIONS

Secure Hash Algorithm (SHA) values are increasingly being used as signatures for identifying objects and ensuring integrity of data segments. The popular Version control system, Git heavily relies on SHA values (pronounced as "sha") to store data and identify data later. Hash is a data structure that maps values to their identifiers, much like labels identify packages. A hash function is a mathematical function that generates a fixed-size identifier from data of any arbitrary size. A simple example is the function that maps full names to initials comprising the first letters of the first and last names. The full names can be of any arbitrary length, but the hash function always generates only two letters.

Cryptographic hash functions impose few more constraints on the hash function. Most important is the collision resistance property. When the hash function returns the same identifier for two different data segments, we say that there is a hash collision. For instance, in the previous example, the hash values for Vishnu Pendyala and Victor Peng collide. Cryptographic hash functions ensure that collisions are extremely rare. For instance, using the SHA-1 function published by the National Institute of Standards and Technology (NIST), the chance of a collision is one in 2^{80} – it takes that many brute force iterations and therefore a lot of computation power at a huge monetary cost, to produce one collision.

Another important property of the Cryptographic hash functions is that even a very minor change to the data results in a huge change in the hash value, so much so that there is no resemblance or correlation between the before and after values. This and other properties of the Cryptographic hash functions make it almost impossible to generate the original data using the hash value, and even minor tampering of the original data will make the hash value completely unrecognizable. The hash value therefore serves as a true and unique signature of the data it represents.

It can be easily seen how Cryptographic hash functions help in the implementation of Blockchain. When the blocks of data are linked by their signature hash values, tampering of the data is not possible without changing all the signatures of all the blocks in the chain.

From the above requirements, it is clear that we need a data structure that is like a linked list, which must be tamper-proof. This data structure is the blockchain. It is a chain of blocks of data, recorded in chronological order, sealed in time, using cryptographic hash functions. It serves as the ledger to keep track of ongoing transactions. The data structure is shared on a Peer-to-Peer (P2P) network, without a single point of failure. Therefore, the data in a blockchain can be audited by any entity on the P2P network. The entity can be an autonomous agent. The data structure can be used to track anything that is worth tracking and not just money or assets, although its role can make much more impact in financial applications. Blockchain can also be viewed as an operating system on which applications like cryptocurrencies run.

The Blockchain constitutes the distributed ledger that is present with all participants. A good analogy for understanding Blockchain is the game of online monopoly. Each player must have a copy of the current state of the game. All players must agree with every move on the board and every transaction. The moves and transactions are unchangeable – the past is immutable. Each remote player has a copy of the board, much like the Blockchain ledger. Each move can trigger a rule, a *smart contract* to execute – smart because the implementation of the contract is automated in software. The smart contract feature can play a key role in the evolution of autonomous systems, which function without any manual intervention.

Note   There are many ways to visualize the Blockchain – as a database, an Operating System, a ledger, a software data structure, a board game.

Genesis Block

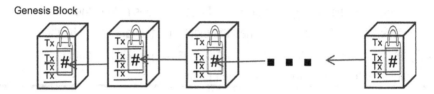

Figure 8-2. *Blockchain schematic*

Though blockchain is used for tracking any entity, let us call the entity being tracked by the blockchain an *asset*. An asset in a blockchain is tracked using its hash value, something like a signature – a unique id. A block is a list of transactions in the ledger along with the cryptographic hash value of the previous block and some other metadata. The schematic is illustrated in Figure 8-2. Each block contains a number of transactions noted as Tx in the figure, sealed by their cryptographic hash value, which is saved in the next block. It is the reference to the hash value of the previous block that gives the chaining effect in a chronological order. The arrows between the blocks represent the references to the previous block's hash value.

The hash value is generated from the content of the block, which is the list of transactions and uniquely identifies the content, as explained in the box titled "Cryptographic Hash Functions". As pointed out there, collisions are extremely rare in SHA-1 and so far have not been found in the SHA-2 set of algorithms. Hence, it is reasonably safe to assume that the hash uniquely represents the information contained in a block. Any small change to the content, such as by way of tampering, will change the hash value substantially. This is called the *avalanche effect*. The first block, called the genesis block, does not have any reference stored, because there is no preceding block.

This simple chaining ensures that the contents of the block cannot be tampered. Any tampering in the information contained in any block will change its hash value substantially. Because this hash value is stored in the next block, the hash value of the next block will also be changed and so on. It is impossible to change the contents of any block without changing every following block in the blockchain. It must also be noted that the cryptographic hash value has another property called *pre – image resistance*, which implies that even if we know the hash value, it is not possible to generate the information that it represents. All these properties of the cryptographic hash values make them ideal for use in the blockchain.

Note   By design, Blockchain becomes increasingly immutable with time and growing number of blocks.

The transactions stored in a block are also represented as a cryptographic hash tree, called *a Merkle tree*. A Merkle tree is a software data structure where records of information, in this case transactions, are stored as the leaf nodes of a graph and interconnected. It is a graph that looks like an inverted tree with the root at the top branching at each level. It is a binary tree with two child nodes under any node and each node being labeled by the hash value of its two children. This cryptographic hash value is a further safeguard against tampering of individual transactions.

Figure 8-3 illustrates a Merkle tree of transactions in a block. The transactions are divided into buckets of two. If the number of transactions is odd, the last transaction is duplicated to make the number even. The hash value of each bucket of two nodes is stored in the parent node. The tree grows bottom up, computing hash values of buckets at each level until we end up in a single root at the top. Figure 8-3 has only four transactions, so three levels are sufficient to construct the Merkle tree. If there are two more transactions, we add another level of nodes containing hash values of the buckets in the lower level. The Merkle tree data structure further helps in verifying the integrity of the transactions by tracing the hash values through the tree.

Note   Cryptography is a common theme across many security
solutions.

Each block contains more metadata in addition to the hash value. It
contains the timestamp, which is the time in seconds the block is created
in the form of the Unix Epoch. By virtue of the cryptographic hash value,
this timestamp becomes immutable as well and further helps to protect
the chronological order from any tampering. The Merkle root hash value at
the top of the Merkle tree of the transactions is also stored in the metadata.
Storing the Merkle root hash in the metadata simplifies the verification
process immensely. This enables *lightweight entities* also to be on the
blockchain P2P network. The lightweight entities need not save the entire
transaction trees, which can run into several gigabytes. They only need to
store the Merkle root. Only full-service entities store the entire Merkle tree.

To verify that a transaction record is in a block, a lightweight entity on
the P2P network only needs the metadata, typically 80 bytes per block,
which includes the Merkle root hash. It can query any full-service node
on the blockchain to return a *Merkle Path* for the transaction, which
comprises the nodes needed to reach the root from the transaction leaf. In
Figure 8-3, to verify that the transaction in the polygon is in the block, the
Merkle Path needed is only the two circled nodes – the one adjacent to the
transaction and the one labeled "Bucket 1 #". If these two nodes are given
by any full-service entity, it can be verified if the Merkle Path indeed leads
to the root in the metadata. In the process, the lightweight node needs to
compute "Bucket 2 #".

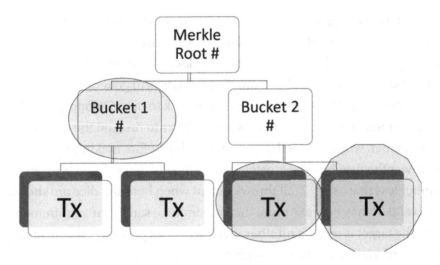

Figure 8-3. *Merkle Tree of Transactions in a block and Merkle Path*

A Merkle tree is extremely useful when the number of transactions increases. In the example in Figure 8-3, where there are only four transactions, we needed two nodes comprising the Merkle Path to get to the root. In general, we need ceil(log_2N) nodes for the Merkle Path, where N is the number of transactions. If we have 16777216 transactions for instance, we need only $log_2 16777216 = 24$ nodes, which is a substantial savings in terms of the network bandwidth, compute power, and storage space.

Note   The Merkle tree is a testimony to the huge difference representation makes and why Formal Methods and Knowledge Representation are key to unfolding promising solutions to the Veracity problem.

Spam emails are sent because often there is not much of a cost to sending spam. To avoid a similar situation with adding blocks to the blockchain, contenders who want to add blocks are required to solve a math puzzle that takes 10 minutes. The puzzle is to generate a hash value that is less than a given value. Going back to our monopoly board game example, it is like asking the online player to first roll the dice to get a number, which is less than a certain value before making a move. The lower the number, the greater is the difficulty. If the target is to get a number less than, say 12, all throws except when both the dice are showing 6 are valid. It may not take more than a minute to satisfy the requirement. But if the target is to get a number that is less than 3, it may take several minutes of trials with the dice to get a 2.

Hash value is unique to a given message. So, how do we generate a hash value that is less than a given value? To generate a new hash value, the information it represents needs to change. The information in a block is changed by adding a *nonce* value. Nonce is a 4-byte number that will change the computed hash value. Because of the avalanche effect of the hash algorithm, a small change to the nonce value causes a huge difference in the cryptographic hash value. True to the English meaning of the word, a nonce is used just once to generate a new value for the cryptographic hash. If the generated value meets the criteria and solves the puzzle, the nonce is recorded in the block header and the block is submitted for inclusion in the chain. However, if the nonce did not help generate a value that meets the criteria, it is discarded and a new value is assigned. Changing the nonce therefore is analogous to rolling the dice to get a new value.

The goal of the puzzle is usually to generate a cryptographic hash value with a number of leading zeros. This is the other way of saying that the generated hash value should be less than a given value. For instance, if the goal is to generate 60 leading zeros in the hexadecimal 32-byte hash value, it can be translated into a constraint that the generated hash value must be less than 0x10000000. The more the number of leading zeros, the more difficult it is to generate it. So, there is a difficulty factor involved, which is also recorded in the block's header. A node that solves the puzzle is said to have generated the *proof – of – work*. The computationally expensive process of solving the puzzle is called *mining*.

The hit-and-trial process of solving the puzzle proceeds in a pretty-much brute force fashion. The nonce value is simply incremented with each iteration and the hash value computed for this new nonce. The iterations stop once the hash value has the required number of leading zeros. The blockchain algorithm makes sure that it always takes 10 minutes to solve the puzzle. This time depends on the computing power and the difficulty factor. As the computing power increases with time, the difficulty factor also needs to increase to keep the time constant at 10 minutes. The difficulty factor is adjusted for every fixed number of blocks.

Note   Blockchain is an agglomeration of multiple well-established techniques to ensure secure transactions that can be entirely trusted in a completely trustless environment.

Blockchain for Big Data Veracity

The biggest contribution of blockchain is the trusted way of tracking the generation of data. The Blockchain technology is particularly useful in the *Internet of Things* (*IoT*) and Big Data era. This can be understood by taking the example of the *connected cars*, which are expected to generate several gigabytes of data every hour. This data is sent to the cloud for processing. Self-driving and connected cars are susceptible to hacking because of their excessive dependence on data. But when every bit of data that is generated by the cars is tracked and made open to audits, the likelihood of hacking greatly reduces or is completely eliminated.

Blockchain can lead to the decentralized control of Big Data sources. Owing to the smart contract feature, there is an increasing likelihood that the sources function in an autonomous way, reducing the scope for subjectivity, human error, and fraud. The P2P network on which the blockchain resides can cross country boundaries and owing to the consensus mechanisms, eliminate country-specific prejudices and tampering. The consensus mechanism makes sure that incorrect blocks cannot get into the system. Any attack can cause denial-of-service, but not be able to tamper with the blocks, particularly the ones far behind, as tampering would mean modifying every block after the one tampered with.

Future Directions

Very few technologies have enabled applications that revolutionized the world. Blockchain is likely to do to the transactions involving entities of value what TCP/IP did to the flow of information. TCP/IP enabled the Internet. Blockchain is enabling trust and veracity. Blockchain is still evolving. Over time, there will be many blockchains in use. The Internet of Things explosion is bound to ensure the widespread use of blockchains. It is imperative that standards be framed for interoperability and implementation. Development requires tooling. APIs and tools need to be extensively provisioned to aid in the process of constructing blockchains.

Currently, the speed at which blocks are added does not seem to scale to the anticipated use of blockchain. For large scale use, the transaction processing speed needs to ramp up substantially. The costs involved too are pretty substantial for wide-spread application. In view of the current costs, it does not make sense to use Blockchain for many things that can benefit from accurate tracking. In spite of all the safeguards and security measures, blockchain is still susceptible to attacks, such as a denial-of-service attack. As blockchain grows in use and evolves, more problems are likely to be uncovered and solved. The biggest impact in the long run is likely to be on Big Data veracity than anything else. More and more sources generating Big Data will join blockchains, improving the quality of data they generate.

Note   Blockchain is a key-enabling technology and an important business driver for the future.

Summary

Blockchain technology implements an excellent mix of security checks for tracking and auditing, which give a huge boost to the veracity aspect of Big Data. In this chapter, we examined a number of techniques that characterize the blockchain and briefly studied how blockchain can help improve the veracity of Big Data. The chronological order, immutability of information, ubiquitous metadata, and smart contracts that implement rules automatically all make the blockchain a promising solution to the problems affecting veracity of Big Data.

EXERCISES

1. If we were to use Blockchain technology for the sole purpose of ensuring data veracity, can we simplify it any further? If we can simplify, can it then be applied to more scenarios than just transactions?

2. Social Media constitute a significant portion of the Big Data. Evaluate the applicability of Blockchain technology to address veracity problems in Social Media.

Index

A

Amazon Mechanical Turk, 40
Artificial neural networks, 111–113
    hidden nodes, 112
Authentic websites, 32
Automated theorem proving
    (ATP), 123
Autonomous systems, 161
Avalanche effect, 162

B

Backtracking, 132
Backward chaining, 131
Bayes theorem, 103
Big Data
    algorithms, 2, 4
    analytics, 1
    areas, 3
    businesses, 1
    characteristics, 9
    computing, 1
    developments, 3
    falsity and inaccuracy, 65
    fission, 3
    fusion, 3
    human, 2–3

identify patterns, 4
Naïve Bayes technique, 4
OSN, 1
overfitting, 4
RDBMS, 5
relationship, V's, 8
sources, 2
support vector machine model, 4
training data, 4
uncertainty, 1
valuation, 7
value, 6
variability, 6
variety, 6, 8
velocity, 6
veracity (*see* Veracity data)
viability, 7
visualization, 7
volume, 5
Web 2.0, 2
Blockchain
    asset, 162
    avalanche effect, 162
    Big Data veracity, 167–168
    concept, 156
    consensus-based truth, 157
    cryptocurrency, 61, 155

N

Printed in the United States
By Bookmasters